Qualitative Analysis
of Physical Problems

Qualitative Analysis
of Physical Problems

M. Gitterman

V. Halpern

DEPARTMENT OF PHYSICS
BAR-ILAN UNIVERSITY
RAMAT-GAN, ISRAEL

 1981

ACADEMIC PRESS

A Subsidiary of Harcourt Brace Jovanovich, Publishers

New York   London   Toronto   Sydney   San Francisco

6474-8212

PHYSICS

Preparation of the manuscript for this volume was made possible through the assistance of Bar-Ilan University, Ramat-Gan, Israel.

ACADEMIC PRESS, INC.
111 Fifth Avenue, New York, New York 10003

United Kingdom Edition published by
ACADEMIC PRESS, INC. (LONDON) LTD.
24/28 Oval Road, London NW1 7DX

Library of Congress Cataloging in Publication Data

Gitterman, M.
  Qualitative analysis of physical problems.

  Includes bibliographical references and index.
  1. Physics––Methodology. 2. Physics––
Mathematical models. I. Halpern, Vivian, Date
joint author. II. Title.
QC6.G54   530'.072   80–25096
ISBN 0–12–285150–1

PRINTED IN THE UNITED STATES OF AMERICA

81 82 83 84   9 8 7 6 5 4 3 2 1

*To the memory of all those who,
by the sacrifice of their lives,
enabled our people
to return to and develop
our country*

Contents†

†An asterisk denotes an advanced topic, as explained in the preface.

vii

Preface

Physical problems that can be solved exactly are the exception rather than the rule. However, physicists and applied mathematicians have developed a number of qualitative methods for analyzing physical problems and obtaining estimates of the magnitude of their solutions. These methods, such as dimensional analysis, symmetry considerations, and perturbation theory, are usually described, often in great detail, in books devoted to each topic separately, or in books devoted to a specific field such as quantum mechanics. The aim of our book, in contrast to this, is to present a general review of the essential features of all the main approaches used for the qualitative analysis of physical problems, and to demonstrate their application to problems from a wide variety of fields. This book is a greatly expanded version of a one-semester course given by one of us (M. G.) to graduate students of physics, and we hope that it, or parts of it, will prove useful, in particular, as a basis for teaching courses.

This book is intended not only for those who wish to learn and apply these methods, such as graduate students, experimental physicists, chemists and engineers, but also for all those who are interested in obtaining a general perspective of modern physics, such as college and high-school teachers. We realize, of course, that most of the methods that we describe will be familiar to experienced theoretical physicists, but hope that even they may find some interest in this review of them.

The level of background knowledge in physics expected of the reader for most of this book is that provided by university courses in general physics, such as the Berkeley or Feynmann physics courses, or by standard intro-ductory textbooks (for instance, those by Halliday and Resnick or by Alonso and Finn). Those few parts of the book that require a wider background knowledge, such as that which a person with a first degree in physics should certainly possess, are indicated by an asterisk. These sections can be omitted without loss of continuity. The knowledge of mathematics that is assumed includes, in particular, familiarity with the standard techniques for solving ordinary and partial differential equations.

We now turn to an outline of the book's contents. Five general principles are relevant to the mathematical formulation and solution of almost all scientific problems, namely (i) the construction of a model, (ii) dimensional analysis, (iii) the symmetry of the problem, (iv) the analytic properties of the physical quantities involved, and (v) the method of the small parameter. These principles will be treated in detail in the body of the book, in Chapters 1–5, respectively, and we restrict ourselves here to a discussion of their importance.

While it is often possible to write down, at least formally, the full set of equations determining the behavior of a given system, an exact solution is frequently not feasible. In that case, the first stage is to simplify the problem as much as possible, a process that is equivalent to replacing the real system by a model one. In addition there are systems, such as the atomic nucleus, for which the appropriate set of equations is not even known, and the most practicable approach is to construct a model and compare its properties with the observed ones of the system. Great care must be taken to ensure that the model is internally self-consistent, and that it corresponds to the real system in any relevant extreme cases. It is not easy to give general rules for the construction of models, but the examples presented in Chapter 1 show various approaches to this problem.

Once a model system has been constructed, it is natural to examine the parameters involved in it and in its solution. Dimensional analysis plays an important role here since only quantities having the same dimensions can be proportional to each other. For instance, the propagation of waves is characterized by the wave velocity v, of dimensions LT^{-1}, while diffusive motion is characterized by a diffusion constant D of dimensions L^2T^{-1}. Hence the distance l over which a disturbance propagates in time t is proportional to vt in the first case and to $\sqrt{Dt}$ in the second. Frequently, a great deal of information can be obtained by scaling the variables so as to obtain a dimensionless equation. Examples of nontrivial applications of dimensional analysis to some typical model systems are presented in Chapter 2.

Another feature of many physical systems from which important general results can be obtained is their symmetry. The classical conservation laws of physics can be derived from the symmetry of space and time, and this provides a deeper insight into their significance. In current theories of the elementary particles, the symmetry-based conservation laws of strangeness, charm, etc., play an important role. The application of symmetry considerations to solids not only simplifies immensely calculations for crystals, but also permits the prediction of which effects can be observed in different types of crystal. For instance, the phenomena of piezomagnetism and magnetoelectricity were first predicted theoretically for certain crystals on the basis of their symmetry and only afterward observed experimentally. These uses of symmetry are described in Chapter 3.

Before attempting to solve a problem, it is also worthwhile to examine the analytical properties of the solution. These are of two types. First, the solution may possess singularities because of symmetry or other physical considerations. It is important to locate such singularities before attempting to solve the problem since solutions may not be well behaved in the vicinity of such points. Second, there frequently exist important connections among different physical quantities because of general principles, such as the causality principle. These connections may not only reduce the amount of work required to solve a problem, but can also impose restrictions on the form of the solution and affect its analyticity. Chapter 4 is devoted to these aspects of the solutions of typical problems.

Once full use has been made of the considerations of the first four chapters, we are ready to start investigating the order of magnitude of the problem's solution. The methods of solution are usually based on the identification of one or more small parameters and the subsequent application of some version of perturbation theory. Such a method of solution assumes that a small change in the parameters produces only a small change in the solution of the problem. That this is not always true is shown, for instance, by the phenomenon of superconductivity. Moreover, there are problems in which no small parameter can be found. For instance, while the ratio of the mean kinetic energy to the mean potential energy per particle is a small parameter for a solid, and its inverse is a small parameter for a gas, this ratio is of order unity for a liquid. The absence of a small parameter in this case is a major difficulty in the development of a microscopic theory of liquids. However, in most cases it is possible to find a small parameter with respect to which some form of perturbation theory can validly be applied. Part of Chapter 5 is devoted to a brief description and comparison of the various forms of perturbation theory. After this, we consider some more difficult situations, such as that in which the small parameter enters as a factor multiplying the highest derivative of a differential equation, or in which only the form of the zeroth-order solution is known. Finally, in Chapter 6, we apply all these methods to a simple example, the calculation of some nonlinear optical properties of a crystal.

The table of contents provides a guide to the topics and problems discussed in each chapter. The first section of Chapters 1–5 is an extensive introduction to the subject matter of that chapter. We have planned these introductions so that they constitute a book within the book, which can in principle be read before the rest of the book or used as the basis for a short course.

The choice of examples to illustrate the general principles is mostly connected with the specific interests of the authors, and so is fairly random. However, an attempt has been made to illustrate the application of the different methods both by simple (and sometimes even trivial) examples, which demonstrate their use clearly, and by more complicated ones, which show the usefulness and power of the methods. Additional examples will be found in the sets of problems at the ends of Chapters 2–5. In order to present some less familiar applications, we have based many of these problems on a reformulation of articles from the *American Journal of Physics*.

The equations and figures are numbered sequentially in each chapter, e.g.,

(1.1), (1.2), etc., and are referred to within the same chapter without the chapter number. An annotated list of references for each chapter is presented at the end of the book. This list is, of course, by no means exhaustive, but rather contains sources that we have found useful.

The choice of units for electromagnetic quantities is, as usual, a difficult problem, since SI units are of great use in practical problems but are not widely used in theoretical physics. We have decided to use SI units in those parts of the book connected with dimensional analysis (Chapter 2) and practical problems such as space-charge-limited currents (Section 1.5). In the rest of the book, we use Gaussian units, i.e., electric charge and electric current are expressed in esu and magnetic fields in emu.

Finally, a few words about the general role of theoretical physics and its place between experimental physics and mathematics: Experimental physics usually takes the results of a number of experiments and uses inductive reasoning to reach conclusions about the causes of the phenomena being measured. In theoretical physics, on the other hand, we normally start from general principles and deduce from these how certain types of system should behave, i.e., we use deductive reasoning. There is an analogy between these two approaches and the relationship between literature and painting. When reading a book, we generally accumulate information and build up a picture of the personalities and situations involved, i.e., we use an inductive approach similar to that of experimental physics. In looking at a painting, on the other hand, we first gain a general impression of the work as a whole, and only then, if we find it interesting, do we proceed to examine the details. This approach is similar to our procedure in theoretical physics. However, while abstract art is not required to bear any obvious relationship to the real world, the axioms of theoretical physics must correspond to the actual physical world, so that only an examination of their practical consequences can convince us of their accuracy and consequent value. This problem of relevance need not concern the mathematician, whose only requirement is that his postulates be logically self-consistent and the results deduced from them be rigorously proved. Moreover, while the mathematician is generally interested in exact results, the theoretical physicist, like the experimentalist, often has to be content with approximate ones. This is why we have concentrated in this book on qualitative and approximate methods.

In a book such as this errors are bound to occur, and we shall welcome being notified of these by readers.

The authors are grateful to Bar-Ilan University for the use of its facilities in the preparation of this book. One of us (M. G.) is indebted to Professor J. Dorfman and Professor R. Zwanzig of the University of Maryland and to Professor I. Oppenheim of the Massachusetts Institute of Technology for their hospitality during the academic year 1977–1978, when part of this book was written.

Chapter 1 | The Construction of Models

1.1 Introduction

The Need for Models

The world around us is full of unsolved problems. These abound not just in the physical sciences, but in all fields in which people attempt to understand and explain what has happened or is happening, and to predict what will happen. For instance, economists want to understand and predict the fluctuations between recessions and booms in the economies of nations, and the effects on investment and growth of changes in the tax laws. Political scientists would like to explain and predict the changes in support for a given policy or political party. An important interdisciplinary problem is the study and prediction of the future of the world's supplies of raw materials and energy resources. In the physical sciences, geologists want to understand the motion of the earth's crust and to predict earthquakes, while meteorologists attempt to explain and predict both long-term climatic variations and the daily variations in the weather. Engineers are concerned with such different problems as control of the motion of rockets in space and the flow of water in sewage systems. And last but not least, physicists are also concerned with an immense variety of problems concerning the behavior and properties of elementary particles, nuclei, atoms, molecules, gases, liquids, solids, plasmas, electromagnetic radiation, gravitation, and so on.

In spite of their great variety, all these problems have one key feature in common. If an investigator, whom we shall call a theoretician, is asked to help solve them, he will at first be horrified at the diversity of problems in the same field. Successive problems, even of the same type, differ from each other in factors, such as the initial conditions, that have a very considerable effect on the solution. Fortunately, the theoretician has a very simple method of making life easier for himself. Instead of solving the problem presented to him, he offers slight variations on the known solution of some different problem. This process is called by the theoretician the construction of a model for the problem. Its success depends on a suitable choice of model. An example of this kind is provided by the motion of a particle under the influence of a restoring force proportional to its displacement from its equilibrium position, i.e., the simple harmonic oscillator, a system for which the exact solution is known. This system has been used by physicists, with a fair degree of success, as a model to help explain such different physical phenomena as the heat capacity of a crystal, the dispersion of light in a medium, and the emission of electromagnetic radiation. Exceptions to the above scheme occur when a physical problem is so novel that it requires the genius of an Einstein to devise an entirely new model, or of an Onsager to solve one exactly. Our book is not intended for this sort of person or problem. If the authors knew a recipe for major scientific discoveries, they would already have used it for their own work. Our aim in this chapter is rather to discuss, with illuminative examples, the considerations that affect the choice of an appropriate model. Our examples here, and throughout this book, will be taken from problems in physics.

Simplification of the Problem

In order to construct or choose a model, it is necessary first to understand the physics of the system under consideration. It is usually worthwhile to start with the simplest possible realistic physical model, in which the geometry is chosen to be as convenient as possible, factors of only minor importance are ignored, and slowly varying factors are treated as constants, for instance. The self-consistency of such simplifications can frequently be checked once the simplest model has been solved: complications will only arise if the solution of the original problem is nonanalytic, a situation that we consider in Chapter 4. Once the model has been chosen, it is necessary to construct the mathematical apparatus to find a solution, such as equations for finding the values of experimental observables. These equations usually contain some parameters, the values of which are determined by comparison with experiment. Because of these adjustable parameters, agreement with experiment does not guarantee the accuracy of a model. Frequently, entirely different models, each with its own parameters, provide equally plausible descriptions of a physical system. In such a case, it is necessary to examine the system in greater detail in order to find features (such as dependence on temperature or pressure or the combined effects of different applied fields and/or temperature gradients) for which the different models predict different results. In this case, the theoretician not only tries to explain what has already been observed, but also suggests new experiments and measurements to test and help develop the theoretical model. All these points are fairly obvious, of course, but the examples in this chapter of how the models are applied to actual problems will, we hope, be of use to the reader.

Microscopic and Macroscopic Approaches

In theoretical physics, two entirely different approaches can be used, and these lead to different types of model. One approach, which we term the microscopic, starts from the forces between the individual components of the system, which may be elementary particles of various types, nuclei, atoms, molecules, or larger bodies according to the type of phenomenon being considered. Alternatively, one can use what we term a macroscopic or phenomenological approach, which is based on some average description of these interactions. The microscopic approach involves mechanics and statistics. The mechanics may be classical, as for large bodies moving slowly, relativistic, as for bodies moving with velocities close to that of light, or quantum mechanical, as for systems at the atomic or subatomic level. In all these cases, its main aim is to find the energies and other characteristics of physical systems at different times. The results obtained from the solutions of such problems are then used in conjunction with statistics to study the behavior of systems at different temperatures. The macroscopic or phenomenological approach, on the other hand, is appropriate for phenomena with a large characteristic scale, so that the details of the atomic and interparticle interactions can be ignored. In this approach, some averaging procedure is used to take account of these interactions. For instance, the average of the mechanical equations of motion for charged particles leads to electrodynamics, and the properties of liquids and gases are studied by hydrodynamics. In all these problems, exact solutions are the exception rather than the rule, so that it is generally necessary to use some approximations. Such approximations amount to the construction of a model of the real system, and it is vitally important to know how relevant is the model and its solution to the experimental results and basic physical laws.

Ideal and Nonideal Gases

The simplest example of a model system is an ideal gas, i.e., a gas of noninteracting particles. This model can be used to describe the properties

of a classical gas at low densities.† The problem of noninteracting particles
can be solved exactly at both the mechanical and the statistical level.‡ As
is well known, statistical calculations lead to the simple connection between
the thermodynamic variables of the system $pV = RT$, where p is the pressure,
V the volume, T the temperature, and R some constant. This equation
describes fairly accurately the properties of the simplest gases. However, it
predicts only a continuous, monotonic change in the system's properties, and
so cannot describe a change from a homogeneous system to an inhomogeneous one, such as the appearance of drops of liquid in a gas. Thus, the model
of noninteracting particles cannot be applicable to phase transitions. The
latter can be described by the more general model of a gas deduced phenomenologically by van der Waals in 1873, for which the equation $pV = RT$ is
replaced by $(p + a/V^2)(V - b) = RT$, where a and b are constants. Such
an equation of state was only later understood to represent a model for a
system of particles with strong short-range repulsive interactions and slowly
decreasing long-range attractive ones. In fact, Kac, Uhlenbeck, and Hemmer
showed that a slight modification of the van der Waals equation provides
the exact solution for a system of hard-core§ particles with attractive forces
of the form $f(r) = a \exp(-ar)$ in the limit $a \to 0$, i.e., infinitely small forces
of infinitely long range. For some temperatures and pressures, the van der
Waals equation has more than one solution for the volume, i.e., it describes
phase separation. Therefore, for a long time it was thought that this model
describes a phase transition quite accurately. However, further theoretical
and experimental developments revealed the principal defects of this model.
The van der Waals equation leads to a negative compressibility, $-dp/dV < 0$,
for some values of the thermodynamic variables. This is in contradiction to
the van Hove theorem, according to which an accurate statistical calculation
can lead only to nonnegative compressibilities, $-dp/dV \geq 0$. Moreover, it
was found that the van der Waals equation disagrees with the experimental
results near a so-called critical point.

The above example shows the usual way in which a model develops.
When it is first presented, the model describes well the main features of the
system. Further investigations, from different aspects, help to clarify the
approximations involved in it. Sometimes, the factors ignored in the model

---

† A counterexample is provided by a gas of electrons at low temperatures, which becomes
more "ideal" as its density increases. This sytem is considered in Section 5.1.

‡ The system of noninteracting particles is an abstraction not only in practice but also
theoretically, since we cannot separate a particle from the field associated with it. If we are
interested in exact solutions, no bodies at all is really too many, since within modern quantum
field theory the problem of zero bodies (vacuum) is insoluble.

§ Hard-core particles are ones that cannot overlap each other; such a requirement is the
simplest example of repulsive forces.

affect the results of more detailed experiments, while on other occasions it may be found to contradict some newly discovered, rigorously accurate, general theoretical results. In both cases, the model has to be refined or to be replaced by a more accurate model.

Systems of Interacting Particles

We now turn briefly to the microscopic theory of a system of interacting particles. There are only four basic types of interaction, namely, the strong, electromagnetic, weak, and gravitational; the strengths of the associated forces are typically in the ratio $1:10^{-2}:10^{-12}:10^{-39}$. We shall not consider gravitational forces, i.e., those attractive forces acting between any pair of bodies and proportional to their masses, since they are dominant mainly in large-scale phenomena such as the motion of celestial bodies. Of the other types of interaction, the electromagnetic is the only one whose value is known exactly. Attractive electromagnetic forces between nuclei and electrons in atoms, molecules, and solids can lead to the appearance of bound states, a feature of crucial importance in all macroscopic phenomena. These electromagnetic forces, in combination with quantum-mechanical effects, determine the chemical structure and mechanical, electrical, and other properties of materials. In the very important special case of solids, we know which particles can combine and the forces between them, so that we can write down the exact quantum-mechanical Hamiltonian for the system. However, a typical solid contains some 10^{23} particles, and the question arises of what to do with such a Hamiltonian, since computers are useless for treating such a large-scale problem. Moreover, the initial conditions are unknown. In order to obtain results, we must construct models that neglect a significant part of the real interactions. For some purposes, detailed calculations on much smaller systems have proved useful. However, it is found that the most effective model of general use is that of elementary excitations, which is now one of the basic methods of modern theoretical physics. As we discuss in Section 1.4, under certain conditions a system of strongly interacting particles can behave as a set of elementary excitations or quasiparticles with scarcely any interaction between them. As a result, the problem can be reduced to a model that has been solved previously, in this case that of an ideal gas.

Examples of the Microscopic Approach

In the first half of this chapter, we describe the microscopic approach to three major physics problems, and the relationship between experiments, models, and theory in each case. We consider first, in Section 1.2, the atomic nucleus, a system for which different types of experimental results led to different, and in many respects contradictory, models. In this case, subsequent theoretical developments of a more refined model suggest that each of these models is, in fact, a valid approximation to the true system for the range of phenomena which it attempts to describe. There is no reason, in fact, that the same simple model should be valid for different types of experimental results, since the approximation involved in deriving a given model may be justified in some cases but not in others.† Then, in Section 1.3, we consider the quark model of elementary particles and the connections between them. This is a system in which experimental results led to a model which predicted more experimental results. When these and related experiments were performed, their results were used to modify and refine the model. This modified model in turn suggests new experiments, and so the process continues. After this, in Section 1.4, we consider the properties of solids, and in particular the electronic properties of metals. For this system, experimental results led to a model (the free-electron model) that was known to be internally not self-consistent. The basic validity of this model was only established by later theoretical developments, i.e., the development of the theory of elementary excitations, for which additional experimental results were not crucial. Thus, there is a complete contrast between elementary particles, for which further experiments are continually required for the development of the theory, and elementary excitations, for which mainly theoretical developments were required. The reason for this difference is, of course, that the forces acting in solids are the well-known electromagnetic ones, while those between elementary particles can at present only be deduced from experiment.

Examples of the Macroscopic Approach

The last two sections of this chapter are devoted, by contrast, to problems treated by the macroscopic approach. While this type of problem may be of less fundamental interest, it is of great practical importance in everyday physics and its applications. Problems of this sort usually involve only a small number of equations, because of the averaging of microscopic interactions that is implicit in the macroscopic approach. The need for suitable models arises when an exact solution of this system of equations cannot be found, as will often happen if the equations are nonlinear or involve partial

---

† A well-known example of this is provided by the dual wave–particle nature of matter, so that electrons, for instance, behave under some circumstances like waves and under others like particles.

differentials. Numerous examples of this sort of system exist. We consider just two fairly typical ones that demonstrate some features with regard to the construction and use of models that are of general applicability. For instance, in both of these problems we consider only the simplest possible geometry since it is essential to obtain and understand the solutions for such cases before examining more complicated systems. In Section 1.5, we consider steady-state space-charge-limited currents in insulators containing electron traps. This is a problem in which we have to examine progressively simpler and simpler models until one or two are found that can be solved exactly. These solvable models prove to be extreme cases that both provide upper and lower bounds on the behavior of the system and also indicate how the system will behave qualitatively in more general cases. Finally, we consider in Section 1.6 the motion past a solid surface of a fluid of low viscosity. In order to solve this problem, it is necessary to divide the fluid conceptually into two regions. One of these is a thin boundary layer near the surface in which the viscosity is dominant and a number of approximations can be made so as to obtain a solvable set of equations. The other region consists of the rest of the fluid, where the effect of viscosity is negligible and the fluid can be treated as ideal, thereby again leading to a solvable set of equations. Thus, this is an example of a macroscopic system in which the physics dictates the use of very different approximations for the same homogeneous fluid in different parts of the system.

Other Applications of Models

Models originally studied for physics problems can also be of use in other fields, including some of those mentioned at the beginning of this section. For instance, the fluctuations in the economy between recession and boom have been compared to the motion of a pendulum, i.e., to simple harmonic motion in which the difference between supply and demand acts as the restoring force. In order to control it, some damping mechanism is required, and this can hopefully be provided by suitable government action. A two-party political system can be compared to an Ising ferromagnet, in which the spin of each atom can point in one of two directions. The nearest-neighbor interactions of the physical system correspond to the influence of friends and colleagues, an external magnetic field to such large-scale effects as the economic situation, the general climate of opinion in the country or political propaganda, and the temperature to the degree of probability associated with each such influence. In other cases, such as the world's supply of essential raw materials, for instance, the construction of a mathematical model is an essential first step, after which all the predictions as to the future can easily be computed. The recent arguments about "dooms-day" have been concerned with the accuracy of the model rather than of the results derived from it. All these examples show how important it is for a theoretician to choose a suitable model, with appropriate values of the parameters, when attempting to solve a problem.

1.2   The Atomic Nucleus

The Need for Nuclear Models

We begin our study of models by considering the atomic nucleus. The need for models here can be understood by contrasting the nucleus with a classical system of charged particles. For the latter, we can choose the number of particles to be studied and their charges as we wish, and measure the forces on each one both when it is at rest and when it is in motion. By means of such measurements, we can derive and test the laws governing the interactions, and these prove to be fairly simple, for instance, Coulomb's inverse square law in electrostatics. When studying the atomic nucleus, on the other hand, we cannot choose freely an aribtrary system of protons and neutrons, at arbitrary distances from each other, and make detailed measurements on each one. Instead, we have to content ourselves with examining the nuclei found in nature, plus a few more than can be produced by suitable reactions. Moreover, the small size and indistinguishability of protons and of neutrons prevent measurements of the detailed forces on a given one, while the com-plicated nature of the forces between these nucleons also prevents a simple analysis. Because of these restrictions, we must just make use of the limited amount of information that can be obtained from the properties of available nuclei, plus their interactions with beams of photons, neutrons, protons, and α-particles, for instance, to describe and attempt to understand these forces. An important first step in such a process is to arrange and classify the ex-perimental results, and it is here that simple models of the system are very valuable. If a set of results can be explained in terms of one such model, it is worthwhile to examine the other properties of the model so as to see how they compare with those of the real system. Any differences that are found place restrictions on the validity of the model, but may also indicate how to im-prove it.

Before describing the essential features of some of the main models of the nucleus, we consider briefly the types of experimental results for which they have to account. The first of these is the mass of the nucleus. It is found that the mass of a nucleus containing A nucleons, of which Z are protons and $N = A - Z$ are neutrons, is less than the sum of the masses of Z free protons and N free neutrons by an amount $\delta M(A, Z)$ which is known as the mass defect of the nucleus. According to Einstein's equation, this mass defect is associated with a binding energy $B(A, Z)$ of the nucleus given by

$$B(A, Z) = c^2\, \delta M(A, Z). \tag{1.1}$$

and the variation of this with A and Z provides some information about the forces between nucleons.

Another important source of information about nuclear forces comes from scattering experiments, in which the scattering by nuclei of a beam of particles is measured. Three types of interaction are possible when a beam of particles of type b is incident on a nucleus X, namely:

(i)   elastic scattering, $b + X \rightarrow X + b$;
(ii) inelastic scattering, $b + X \rightarrow X^* + b$, where X^* denotes an excited state of the nucleus;
(iii) real reaction, $b + X \rightarrow X' + b'$.

The scattering cross section $\sigma(\theta)$ is defined as the ratio of the number of particles emitted per unit solid angle at angle θ to the incident beam per unit time to the number of particles crossing unit area in unit time in the incident beam. Its variation with the angle θ and with the energy of the incident beam can provide important information about the excited states of the nucleus, since inelastic scattering and real reactions are only possible if the incident particle can donate sufficient energy to the nucleus.

The Liquid-Drop Model

Two prominent properties of the nucleus are the approximate constancy of the binding energy per nucleon and that the nuclear radius R, as revealed by the scattering of electrons, is proportional to $A^{1/3}$, except for the very lighest nuclei. Both these properties can be explained if the forces between nucleons are of very short range and the nucleons are packed in the nucleus at a constant density, just like the molecules in a drop of liquid. This analogy with a liquid drop can also explain most of the other terms in von Weiszacker's semiempirical formula for the binding energy of a nucleus

$$B(A, Z) = C_1 - C_2 A^{2/3} - C_3 Z(Z - 1)/A^{1/3}$$
$$- C_{\text{sym}}(N - Z)^2/A + [(-1)^Z + (-1)^N]C_{\text{pair}}\, A^{-3/4}. \tag{1.2}$$

In this formula, the first term arises from the mean binding energy per nucleon inside the nucleus. The second term allows for the fact that nucleons on the surface of the nucleus have less neighbors than those in the center and so are less tightly bound; this effect is analogous to surface tension in a liquid and is proportional to the surface area, i.e., R^2 or $A^{2/3}$. The third term represents the Coulomb repulsion between pairs of protons; the number of such pairs is $\frac{1}{2}Z(Z - 1)$, and the mean inverse distance between the protons is proportional to R^{-1} and so to $A^{-1/3}$. The last two terms cannot be explained by the liquid-drop model and allow for the observed preference for N to be approximately equal to Z and for both of these numbers to be even. Another feature for which the liquid-drop model cannot account is that nuclei are especially stable when they contain the following numbers (known as "magic" numbers) of protons or neutrons: 2, 8, 20, 28, 50, 82, 126.

According to the liquid-drop model, the nucleons are expected to move around very rapidly within the nucleus, with frequent collisions, just like the molecules in a drop of liquid. The excited states of such a liquid drop will involve distortions and oscillatory motions of the drop, with the most important oscillations corresponding to surface waves. The energies associated with such waves are calculated to be several MeV above the ground state energy, and many excited levels are indeed found in this region. However, there are also excitations at much lower energies, and this sort of model can only account for them if it is assumed that the drop is spheroidal rather than spherical, with these excitations corresponding to rotations about the axis of symmetry.

Another type of experimental result that can be accounted for by the liquid-drop model is the instability of very heavy nuclei. If a nucleus is distorted from a spherical to an elongated shape, its surface area increases, but the average distance between the protons also increases. Thus, the magnitude of the second term in Eq. (2) for $B(A, Z)$ increases, while that of the third term decreases. For light nuclei, the net result is a decrease in $B(A, Z)$, so that the spherical shape is the most stable. However, since the surface energy is proportional to $A^{2/3}$ while the Coulomb term is approximately proportional to $Z^2 A^{-1/3}$, as the atomic number Z increases the effect of the Coulomb term becomes more important, and for very heavy nuclei it may dominate. In that case, the ground state of the nucleus will no longer be spherical, and the spontaneous deformation, or that induced by the capture of an incident particle such as a neutron, may be large enough to make the nucleus unstable with respect to fission.

The Shell Model

In spite of its successes, the liquid-drop model cannot account for a number of experimental results, such as the magic numbers and the preference for N to be equal to Z and for both to be even. In addition, the model of nucleons in rapid motion and colliding with each other ignores the fact that nucleons, like electrons, have spin $\frac{1}{2}$ and so obey Fermi–Dirac statistics and are subject to the Pauli exclusion principle. An alternative model of the nucleus, which does not suffer from these defects, is the shell model, analogous to the atom with its shells of electrons. In the shell model, each nucleon is regarded as moving independently, but subject to the exclusion principle, in a smoothly varying mean potential due to all the other nucleons. Unlike in the atom, this potential will not have a deep minimum at one particular point. Instead, a potential of approximately rectangular form, together with an appropriate strong spin–orbit interaction, leads to energy levels with large gaps for the addition of an extra nucleon beyond the magic numbers. From the known nature of nuclear forces, it is also expected that the differences between the actual forces and the mean potential will cause like nucleons in different levels to pair off with one another so that their total angular momentum is zero, and this explains the greater abundance of nuclei with even values of Z and N. This shell model also accounts well for the observed angular momenta of nuclei, and quite well for their magnetic dipole moments, but not for their electric quadrapole moments. With an appropriate choice of the potential, it can account well for the energies of individual nucleon levels, as found from experiments in which a nucleon is removed from or added to a nucleus. For all these properties, the shell model of the nucleus is far superior to the liquid-drop model; however, it cannot readily account for the total energy of nuclei or for nuclear fission, the phenomena for which the liquid-drop model was proposed, nor for many of the observed excited levels.

Compound Nucleus and Optical Models

One source of additional information about the nucleus is the way in which it scatters an incident beam of neutrons. Here too, just as for the properties of unperturbed nuclei, different experimental results can most readily be described by two very different models. In some cases, a large number of narrow resonances (i.e., peaks in the scattering cross section at well-defined energies) fairly close to each other in energy are observed. This result can most readily be explained in terms of the compound nucleus model. The basic idea of this is that a many-nucleon system can have very many different configurations with only small energy differences between them. While the lowest levels will be perfectly stable, the higher ones will become unstable against particle emission. However, this particle emission will in general be a slow process, since the available energy is mostly shared equally by all the nucleons, and only occasionally will a single particle acquire enough energy to escape. The resonances in the scattering cross section occur because neutrons can only be captured by a nucleus if the total energy of the resulting system corresponds to an excited state of the compound nucleus. If this does happen, there is a possibility of inelastic scattering, since an emitted particle will usually leave the nucleus in an excited state, or of a real reaction. However, this model cannot account for all the observed experimental data on scattering. In other experiments, especially at higher energies, the incident beam of neutrons is scattered in a manner analogous to the scattering of a beam of light by a strongly dispersive medium. Such scattering is most easily explained by the optical model, in which the nucleus is treated as a potential well through which neutrons can pass without colliding with any nucleons.

Use of Conflicting Simple Models

An examination of the above models shows that they are not only quite different but also are not mutually consistent. For instance, the liquid-drop model regards the nucleons as moving around in the nucleus with only a very short mean free path between collisions, so that any excess energy introduced into the system is rapidly shared between them. According to the simple shell model, on the other hand, the nucleons are virtually independent, and can only be excited individually into one of the unoccupied levels. The optical model is certainly not consistent with the liquid-drop model, and on the simple shell model it is not obvious why the strong interactions associated with the incident neutron should not profoundly modify the mean nuclear potential well. Only the recent development of a unified model of the nucleus, based on a shell model with collective excitations of the nucleons, enables us to account with a single model for most of these properties.

In spite of the contradictions between them, all of the simple models of the nucleus that we have mentioned are useful for describing some sets of phenomena. This is a very valuable function of models for systems where no exact theory exsists. A classification of the properties of a system according to such models provides a simple way of describing them, and saves us from having to deal with a large number of seemingly unconnected experimental results. There is no reason whatsoever to require such models to be consistent with each other, nor even to take account of known features of the system, just as the liquid-drop model of the nucleus ignore the fact that nucleons obey Fermi–Dirac statistics. Such features only have to be included when one tries to construct a more refined model of the system, for which the different simple models are just limiting cases valid for the phenomena they describe.

1.3 The Quark Model of Elementary Particles

On 12 and 13 November 1974, the editors of *Physical Review Letters* received two articles announcing the discovery in Brookhaven National Laboratory and in Stanford Linear Acceleration Center of a new elementary particle. Three weeks later, Italian physicists from Frascati telephoned news of the discovery of the same particle to the editors of *Physical Review Letters*. On 2 December 1974, three articles, one after another, were published in this journal. Two years later, the American physicists Samuel C. Ting and Burton Richter received the Nobel prize for "their pioneering work in the discovery of a heavy elementary particle of a new kind."

Definition of Elementary Particles

In order to understand the importance of these results we have to examine the development of knowledge about the elementary particles and of the models used for their description. The meaning of the concept "elementary" changed as our knowledge developed. There was a time when atoms were considered as "elementary particles," but at the beginning of the 1930s four elementary particles (electron, proton, neutron, and γ-quantum) were known. After this, the number of new particles started to grow like an avalanche, with the discovery of the μ-meson in 1936, the π-meson in 1947, strange particles in the fifties, and resonance particles in the sixties. The total number of particles exceeds a hundred and, in addition, for each particle there exists an antiparticle with opposite values of all its quantum numbers. Apart from the electron, proton, neutrino, γ-quantum, and their antiparticles, all the other elementary particles have short lifetimes and decay into other particles. It might seem reasonable to assume that from a study of the decay of a particle one can recognize its "structure," and so gradually arrive at the concept of elementariness. However, in experiments on the β-decay of nuclei, electrons appear which did not exist in the nucleus. Such a phenomenon is also inherent in some other decay processes. Therefore we cannot readily recognize the "internal structure" of the decaying particle and hence decide whether the particle is "elementary" in the usual meaning of this concept, just by examining its decay. Nevertheless, all particles are now believed to be made up of a few "elementary" ones. Modern physics came to this conclusion after a long and very interesting process of development, in which the models stimulated new experimental study, which, in turn, resulted in an improvement of the old models and the introduction of new ones.

Classification of Particles

The classification of the elementary particles must serve as a basis for the construction of models. In other words, we have to look for regularities of any kind, and then construct a model to explain them. Thus, the first question is what regularity can be found.

The experimental results show that all the particles can be divided into four groups of increasing mass; the photon, leptons, mesons, and baryons. Photons take part only in the electromagnetic interactions, while leptons (electron, muon, and two kinds of neutrino) are involved in the electromagnetic and weak interactions. Mesons and baryons, which combine into so-called hadrons, are subject to the strong force, as well as the electromagnetic and weak forces.

All baryons, the most important of which are protons and neutrons, have a half-integral spin and are therefore fermions. The mesons, on the other hand, are described by Bose–Einstein statistics and have a spin 0 or 1. The hadrons can be characterized by a so-called baryon quantum number (the baryon charge), which is equal to 1 for the baryons, -1 for the antibaryons and 0 for the mesons. The processes of creation, interaction, and decay can be formulated as a conservation law for the baryon charge. From this it follows, for example, that a baryon has to be among the products of baryon decay, or that a baryon can appear only together with an antibaryon when the total baryon charge is equal to zero, and so on.

* Symmetry Groupings

Conservation laws, as we shall see in Chapter 3, are not only of practical convenience but also correspond to some symmetry properties of the system. If we know that a system possesses a certain symmetry, we can derive the associated conservation law of the relevant quantum numbers. Unfortunately, however, we do not know the forces of interaction between elementary particles, and so, for instance, we do not yet know which symmetry properties are reflected by the law of conservation of the baryon charge. Instead, we have to solve the inverse of the usual problem, and attempt to find a symmetry that leads to the observed conservation laws. In order to do this, we first have to study phenomenologically the conservation laws and symmetry properties of elementary particles, and then attempt to distribute the particles according to the rules of symmetry theory. The use of the theory of symmetry, with its stringent system of classification, facilitates the systematization of experimental data, the construction of models which then stimulate new experiments, and finally, the identification of the forces acting between the elementary particles.

The existence of common properties for some groups of elementary particles forms the basis of the symmetry classification. In the framework of so-called unitary symmetry, these groups are designated as supermultiplets. We will restrict our attention to the special unitary group of square matrices of order 3, the group SU(3), which has been used by Gellman and Neeman for the classification of elementary particles. Particles having the same isotopic spin and strangeness are placed in the same supermultiplet. The isotopic spin is determined by the number of particles in the supermultiplet, and is defined by the rule that the number of particles in a supermultiplet is one more than twice the isotopic spin. Strangeness measures the distribution of charge among the particles, and is equal to twice the average charge minus the baryon number.† It can be shown that all the particles in the same supermultiplet have the same spin, and also the same parity, i.e., symmetry with respect to inversion. The number of particles in a supermultiplet must equal the dimension of an irreducible representation of the group, so that for SU(3) the possible values are 1, 3, 6, 8, 10, 15, 21, 24, 27, ...

A careful analysis of the experimental data showed that if the elementary particles are grouped together according to their properties, they are in fact arranged according to the above-mentioned supermultiplets. It was found after this classification that in one supermultiplet, one place was unoccupied. All the properties of the hypothetical particle needed to complete the supermultiplet were determined by its place there. The experimental discovery of this particle, the Ω-hyperon, in 1964, was the greatest success of the theory of unitary symmetry. However, two facts indicate the approximate character and incompleteness of the theory. First, it turned out that the particles of a given multiplet differ slightly in their masses. This fact contradicts the demands of theory and makes it an approximate one. Second, certain quantum numbers are conserved only in some interactions. This means that the appropriate conservation laws are "approximate," unlike the "exact" conservation of electron or baryon charges. For example, some of the quantum numbers are conserved in strong interactions, but not in the weak ones.

The Quark Model

A very surprising feature was that all the known elementary particles fall into supermultiplets containing 1, 8, or 10 particles, and there are no particles in the multiplets with dimensionality 3, 6, 15, 21, etc. This fact

---

† The strangeness quantum number vanishes for all hadrons, except for the so-called strange ones, which have anomalously long lifetimes compared with other hadrons (10^{-7}–10^{-10} sec, as compared to 10^{-20}–10^{-23} sec).

Table 1

Properties of the Quarks

	Mass (GeV)	Electric charge	Spin	Baryon charge	Strangeness
d	0.336	$-\frac{1}{3}$	$\frac{1}{2}$	$\frac{1}{3}$	0
n	0.338	$\frac{2}{3}$	$\frac{1}{2}$	$\frac{1}{3}$	0
s	0.540	$-\frac{1}{3}$	$\frac{1}{2}$	$\frac{1}{3}$	-1

suggested to Gellman and Zweig the idea of the following model of elementary particles. All hadrons are constructed from more fundamental particles, named "quarks," which belong to the supermultiplet of dimensionality 3. It is worth noting that the quarks have fractional electric (and baryon) charge, in contrast to all known elementary particles. Table 1 gives the properties of n-, d-, and s-quarks; the antiquarks have the opposite values of all quantum numbers.

According to this theory, mesons consist of a quark and antiquark. In this way, we find nine different combinations divided into two groups with eight particles and one particle, respectively. Baryons consist of three quarks and antibaryons of three antiquarks. The 27 different combinations arising in this way are divided into two groups with eight particles, one with ten, and one with just one particle. It is easily shown that the quantum numbers of these sets of two and three quarks agree with the experimentally found properties of mesons and baryons, respectively. For example, the spin of an even number of quarks is an integer, as it has to be for mesons, and the spin is half-integral for the odd number of quarks composing the baryons. Furthermore, every allowed combination of quarks yields a known particle.

Modifications of the Quark Model

The quark model explains beautifully the distribution of particles over the supermultiplets of the group SU(3), but this model, in turn, raised additional questions, which gave rise to the hypothesis of colored and charmed quarks. The basic problem is that quarks, as elementary particles, have to obey certain statistics. Since they have half-integral spin, they must be described by Fermi–Dirac statistics. However, the Pauli exclusion principle forbids two Fermi-particles having the same quantum numbers. Therefore, it is not clear how three identical quarks can occur in a baryon. This general result, together with some experimental data, led to the model of colored quarks. According to this model, each quark has an additional quantum number called color, which can assume one of three different values. Replacing the group SU(3) by the new symmetry group, the color group SU(3), should not destroy the success achieved by group SU(3) in explaining the formation of elementary particles from quarks. However, the following two additional quark rules have to be adopted. Baryons (and antibaryons) must be made up of three quarks, all of which have different colors. Mesons are made up of a quark and an antiquark of the same color, but with equal representation of each of the three colors. Thus, quarks have a color, but hadrons have not, i.e., they are colorless. In this way, the introduction of color into the quark model enables us to eliminate the contradiction with Fermi statistics, without affecting the success of the model in regarding the hadrons as made up of quarks.

A quantitative criterion supporting the colored quarks model appears when we examine the process of annihilation of electrons and positrons at high energy. From the analysis of some experiments, it follows that at high energies (about 2 GeV for the total energy of the colliding electron and positron) quarks behave as independent particles like leptons. Therefore, the probability of producing hadrons should vary with energy in the same way as the probability of producing a pair of muons. This means that the ratio of hadrons to muon–antimuon pairs $(\mu^+\mu^-)$ should be a constant, independent of the collision energy and equal to the sum of the squared quark charges. Thus we should find that

$$R = \frac{\sigma(e^+e^- \to \text{hadrons})}{\sigma(e^+e^- \to \mu^+\mu^-)} = \sum_i Q_i^2, \qquad (1.3)$$

where σ is the cross section for a process. The initial formulation of the quark model predicted $R = (\frac{2}{3})^2 + (\frac{1}{3})^2 + (-\frac{1}{3})^2 = \frac{2}{3}$, because the index i in Eq. (3) assumes three values, equal to the number of quarks. If each quark is produced in three different colors, we expect $R = 2$. Experimentally, at energies between 2.5 and 3.5 GeV we find $R = 2.5 \pm 0.5$, which strongly supports the existence of three colors and is inconsistent with colorless quarks.

Another important modification of the quark model, in addition to the introduction of color, is the charm hypothesis. The earliest motivation for this was based on an analogy between quarks and leptons. Just as there are four known leptons, so the existence of a fourth quark, designated c (charm) was postulated. The c-quark has the same quantum numbers as a n-quark, except for a value + 1 for the new quantum number charm, and a mass of 1.5 GeV. All the other quarks have zero charm. Later on, the charmed quark was invoked to explain the suppression of certain particle decays that are expected on the three-quark model but are not observed.

Experimental Confirmation and Outstanding Problems

We can now return to the discovery mentioned at the beginning of this section. The new particle found in 1974, called a ψ-particle, is the meson made up of a c-quark and a c-antiquark. Full confidence in this hypothesis came with the discovery shortly afterwards of the ψ'-particle, and in May–June 1976 of the D^0- and D^+-mesons. These three particles are all excited states of the ψ-particle. By now, there seems no doubt that charm and color are found in experiments, and that the new particle found in 1974 is a straightforward proof of the quark model.

Why is the confirmation by experiment of the quark model so important? Quarks are distinguished from leptons by color and also by participating in the strong interactions. It is therefore natural to assume that color is connected with the strong interaction. We can say that the color is the "charge" for the strong interaction, just as the electric charge is responsible for the electromagnetic interaction. From the color SU(3) theory, we find that the quark–quark interaction is mediated by massless particles called gluons, just as the photon is the carrier of electromagnetic forces. When a quark emits or absorbs a gluon, it changes only its color, and stays a quark of the same kind. Like photons, gluons are electrically neutral and have a spin of 1. If we recall Yukawa's assumption that the weak interaction is mediated by the vector π-meson, we come to the conclusion that the strong, weak, and electromagnetic forces are all carried by the same kind of particles. This could be an important stage on the way to the achievement of some simple unified theory.

A step towards such a unification would be the Salam–Weinberg theory of reconciliation of the weak and electromagnetic interactions at high (about 100 GeV) energies, which in many ways resembles Einstein's theory of gravitation. The role of gravitation in that theory is played here by the colored quark model for the strong force. It would be very exciting if all the forces known in nature could be explained by the same kind of theory and then combined into some unified theory.

Returning to the quark model, we must mention that there are a number of unresolved questions for which a new explanation, and perhaps even a new model, is needed. First, we have no satisfactory explanation at present of the anomalously long lifetimes of some of the new particles, nor for the gradual increase in the value of R, defined by Eq. (3), to about 5 at energies of 5 GeV.† Moreover, there is no simple explanation of why individual quarks have not been observed. Another question is why the universe, including the sun and the other stars, is built up only from four of the elemen-

---

† The model of four quarks with three colors gives $R = 3[(\frac{2}{3})^2 + (\frac{1}{3})^2 + (-\frac{1}{3})^2 + (\frac{2}{3})^2] = \frac{10}{3}$.

tary particles (electron, electron neutrino, u- and d-quarks), while the other four (muon, muon neutrino, s- and c-quarks) appear only at high energies. Thus, as we can see, a lot remains to be done in elementary particle physics before a completely satisfactory model is derived.

1.4   Elementary Excitations in Solids

We now turn to a system that is the complete opposite of the atomic nucleus or the elementary particles, namely, the properties of a crystalline solid. For this system, the nature of the particles present (electrons and ions or nuclei) and their interactions are known, but difficulties arise because of the vast number of them present in a typical crystal. We have seen in Section 1 that the simplest model of a many-body system, namely, the ideal gas, does describe some of the properties of real gases. The question that arises is, how we can treat solids? Is it also possible to apply to them this model, which is practically the only one that we can solve exactly? At low temperatures, solids in thermal equilibrium have a crystalline structure. As the temperature increases, the atoms start to oscillate about their equilibrium positions, so that inhomogeneities appear in the distribution of atoms in the lattice. However, these inhomogeneities remain small up to the melting point in most crystals. As a result, it proves possible to describe these inhomogeneities by the model of an ideal gas of "particles" that are called phonons, as we show later in this section. However, the electronic properties of solids, and especially of metals, present a much more challenging problem, which we shall now describe.

The Free-Electron Model

At least for metals, qualitative physical considerations make it very difficult to see how the model of an ideal gas could be applicable to electrons. For an ideal gas, no interactions exist between the different particles, and only their kinetic energy is of importance. In metals, on the other hand, the electrostatic interactions between the electrons can certainly not be regarded as small. In fact, at the typical electron concentration n of 10^{23} cm^{-3}, the ratio of the kinetic energy of an electron, $\varepsilon_k \sim \hbar^2 n^{2/3}/2m$, to its potential energy, $\varepsilon_p \sim e^2 n^{1/3}$, is of order unity. Nevertheless, at the beginning of this century, the model of free electrons was applied to the description of the electronic porperties of metals. Even more strangely, this model (Drude's model) described quite well a number of the electronic properties of metals, such as the temperature dependence of their electrical and thermal conductivites, and the paramagnetism of the alkali metals. The free-electron theory leads to a universal ratio of the coefficient of heat conductivity κ to the electrical conductivity σ. This ratio is described by the Wiedemann–Franz law, $\kappa/\sigma = 3(k_B^2/e^2)T$, where e is the electronic charge, k_B the Boltzmann constant, and T the temperature, a result in quite good agreement with experiment. Even at that time, the limitations of the model were obvious; for instance, it contradicts the results obtained from measurements of the electronic specific heat. This fact, by itself, does not necessarily invalidate the model, since as we saw in Section 1.2 a model may have only a restricted field of applicability. However, the agreement between theory and experiment deteriorated when Lorenz, in 1905, improved the Drude model by assuming a Maxwellian distribution for the electron velocities instead of assuming that they all have the same velocity. This improvement of the theory led to a factor of 2 instead of 3 in the Wiedemann–Franz law, in contradiction with the experimental results. If an improvement of the model worsens the agreement with experiment, there must be something basically wrong with the model. In spite of this, the free-electron model, which is deliberately not based on physical arguments, does lead to quite a number of correct results.†️ This seems very paradoxical, and an explanation is demanded of why such a bad model leads to such comparatively good results.

Normal Coordinates

In order to answer this question, we must first understand how, in general, the properties of an ideal gas can occur in a system of interacting particles. To do this, let us examine the problem of normal coordinates for one of the simplest examples of such a system, a linear chain of identical particles oscillating about their equilibrium positions as shown in Fig. 1. We assume that in equilibrium the particles are located at distance a one from another, and denote by u_n the (small) displacement from its equilibrium position of the nth atom from the origin, in a direction perpendicular to the chain. The potential energy V of the system of particles in a given set of positions can be expanded as a series in the u_n in accordance with Taylor's theorem:

$$V = V_0 + \sum_n u_n \left(\frac{\partial V}{\partial u_n}\right)_0 + \sum_n \sum_{n'} u_n u_{n'} \left(\frac{\partial^2 V}{\partial u_n \, \partial u_{n'}}\right)_0 + \cdots. \qquad (1.4)$$

---

† We note that the problem of superconductivity did not exist at that time, and ferromagnetism was not associated with the electronic structure of the solid. The free-electron model cannot, of course, be applied to the explanation of these phenomena, as they are completely dependent on the interactions between electrons.

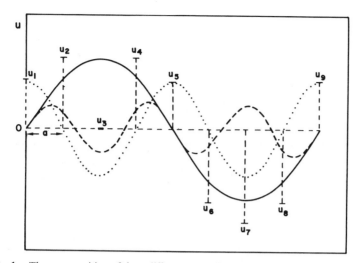

Fig. 1. The superposition of three different normal modes of motion of a linear chain at a given instant, leading to the displacements $u_1 - u_9$ of the particles.

The linear term in Eq. (4) vanishes, since the sum of all forces acting on the nth particle, $-\partial V/\partial u_n$ must equal zero in equilibrium. For an infinite chain, the coefficients $(\partial^2 V/\partial u_n \, \partial u_{n'})_0$ depend only on the distance between particles n and n', so that

$$\left(\frac{\partial^2 V}{\partial u_n \, \partial u_{n'}}\right)_0 = f(n - n'). \tag{1.5}$$

If, for all n, $u_n = \text{const} = b$ say, i.e., all the particles are shifted through a distance b, the resulting forces acting on a given particle vanish, and so

$$-b \sum_{n'} f(n - n') = 0. \tag{1.6}$$

From Eq. (4), we obtain the system of equations of motion

$$m\frac{d^2 u_n}{dt^2} = -\sum_{n'} f(n - n')u_{n'}. \tag{1.7}$$

Let us seek a solution of Eq. (7) of the form

$$u_n = A \exp(iqna - i\omega t) + A^* \exp(-iqna + i\omega t). \tag{1.8}$$

Such a function is a solution if

$$m\omega^2 = \sum_{n} f(n) \cos(qna) \tag{1.9}$$

as can be seen by substituting from (8) into (7) and making use of (6).

The values of the wave vectors q depend on the boundary conditions imposed on the chain. For the sake of convenience, we assume that the chain closes on itself, so that particle $N + n$ coincides with particle n, i.e., $u_{n+N} \equiv u_n$ (periodic or cyclic boundary conditions). In this case, the possible values of q are the N numbers

$$q_r = (2\pi/a)(r/N), \qquad r = 1, 2, \ldots, N. \qquad (1.10)$$

As is well-known from the theory of differential equations, the general solution of the system of equations (7) can be written as a linear combination of the partial solutions (8). The partial waves in such a sum will, in general, have different wave numbers q, frequencies ω_q, and amplitudes A_q, so that the general solution of (7) is of the form

$$u_n = \sum_q [A_q \exp(iqan - i\omega_q t) + A_q^* \exp(-iqan + i\omega_q t)]$$

$$= \sum_q [a_q \exp(iqan) + a_q^* \exp(-iqan)], \qquad (1.11)$$

where $a_q = A_q \exp(-i\omega_q t)$.

The coefficients a_q, or to be more precise simple combinations of them, are called the normal coordinates, since the potential and kinetic energies of the system are the sums of the squares of these coordinates and of appropriate corresponding generalized momenta. These normal coordinates are the solutions of the simple harmonic oscillator equations.

Quasi-particles

The normal coordinates a_q do not depend on which particle n in the chain one is considering; i.e., Eq. (11) determines the displacement of any atom as a sum of the same normal coordinates (with the appropriate phase factors), as shown in Fig. 1. Thus, the oscillations of a system of interacting particles can be described by a set of normal coordinates, provided that the displacements of the particles are small enough for terms of higher than second order in u_n in Eq. (4) (i.e., in the Taylor series for the potential energy V) to be neglected. This independence of the normal coordinates means, in particular, that with a suitable choice of the initial conditions it is possible to excite only a single normal coordinate, so that all the particles oscillate with the same frequency, the value of which depends on the properties of the system. In quantum-mechanical terms, to each normal coordinate corresponds a sound quantum, or phonon, with quasi-momentum $\hbar q$ and energy $\hbar\omega_q$. These phonons do not interact with each other, and so can be treated as an ideal gas, as we mentioned at the beginning of this section.

Thus, the small oscillations of the interacting particles can be regarded as an ideal gas of some quasi-particles, namely, phonons. It is natural to call them "quasi-particles" or "elementary excitations," since they differ from the particles composing the original system. The approximation represented by Eq. (4)–(11) can be written symbolically in the following way:

$$H = \sum_k H_k + \sum_k \sum_l H_{kl} \rightarrow H' = \sum_\alpha H'_\alpha + \gamma \sum_\alpha \sum_\beta H'_{\alpha\beta}, \qquad \gamma \ll 1. \quad (1.12)$$

Here, H_k and H_{kl} are, respectively, the energy (Hamiltonian) of a given particle and the interaction energy between the particles. By means of some transformation, which in our example was the introduction of normal coordinates, this Hamiltonian H can be reduced to a Hamiltonian H', which is the sum of the energies H'_α of quasi-particles and small interaction energies $\gamma H'_{\alpha\beta}$. The interactions between these quasi-particles can be neglected when $\gamma \ll 1$. In such a case, the original problem of a system of interacting particles is replaced by a new problem involving an ideal gas of quasi-particles. These quasi-particles are not like the original particles, although their properties depend on the interactions between the original particles.

The Successes and Failures of the Free-Electron Model

We now return to the paradox of the unexpected success of the theory of free electrons. This can be explained if we assume that the Hamiltonian for the real, strongly interacting, electrons can be transformed according to Eq. (12). In other words, experiment supported not the theory of free electrons but that of some quasi-particles whose properties correspond to those of the given system of electrons. A more careful examination shows that the theory of free electrons describes in a qualitatively correct way those properties that are determined by the statistical properties of the many-body system. According to quantum mechanics, electrons obey Fermi–Dirac statistics, i.e., their equilibrium distribution function at temperature T is

$$f(\varepsilon) = \left[\exp\left(\frac{\varepsilon - \mu(T)}{k_B T}\right) + 1 \right]^{-1}, \quad (1.13)$$

where ε is the electron energy and $\mu(T)$ the chemical potential at temperature T. When $T = 0$, $f(\varepsilon)$ is unity for $\varepsilon < \mu(0)$ and zero for $\varepsilon > \mu(0)$, as shown by the full line in Fig. 2. The value of the chemical potential at temperature $T = 0$ is called the Fermi level, $E_F = \mu(0)$. Thus, at $T = 0$ all states with energy less than E_F are full, and all those with energy more than E_F are empty. As a result of the Pauli exclusion principle, only one fermion (i.e., a particle obeying Fermi–Dirac statistics) can occupy each state. Because of

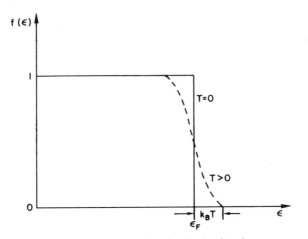

Fig. 2.   The Fermi–Dirac distribution function.

this, as the temperature is raised only electrons from a region of order $k_B T$ below the Fermi level can be thermally excited into states with energies above E_F, so that the distribution function becomes as depicted by the dotted line in Fig. 2. Thus, at temperature $T \neq 0$, only a small fraction $k_B T/E_F$ of the total possible number of excitations appears, so that the system is indeed only slightly perturbed from its ground state.

However, this explanation of the success of the free-electron model also explains one of its failures. If at temperature T the number of elementary excitations that appears is proportional to T and the average energy of each is of order $k_B T$, the total energy of the system increases as T^2. Hence, the electronic specific heat is proportional to T, in agreement with experiment but in contradiction to the Drude model of an ideal gas of electrons. Although the electrons are in fact still an ideal gas without interactions, the change from the classical statistics of the Drude model to Fermi–Dirac statistics and the associated Pauli exclusion principle can be regarded as introducing an "effective interaction" between them. This is responsible for the fact that only a small fraction of the electrons can participate in different kinetic processes. Moreover, a quantitative description of experimental results, for which numerical values of particle parameters such as masses, concentrations, etc., are required, cannot be provided satisfactorily by the theory of free electrons. In order to explain some of these properties, it is even necessary to assume that the quasi-particles have a special dispersion law, i.e., connection between the wave vector q and energy ε_q, very different from the form $\varepsilon_q = \hbar^2 q^2/2m$ that applies to free electrons.

One of the parameters of the theory of free electrons is their mean free path λ. This is introduced to describe the kinetic properties of electrons in a manner similar to the kinetic theory of gases. Comparison of the calculated electrical conductivity and that measured experimentally showed that λ is frequently of the order of hundreds of angstroms, i.e., many times the inter-atomic distance in the crystal through which the electron moves. This result is very hard to understand in terms of the classical theory of free electrons, according to which we would not expect the electron to travel more than one or two interatomic distances without collisions. However, as we have just shown, the theory of free electrons really refers to free quasi-particles. These, like the normal modes of a mechanical system, can extend over the whole solid, so that there is no reason whatsoever why their mean free path should not extend over hundreds of interatomic distances.

*Magnetic Properties of the Electron Gas

Another feature of an electron gas is its paramagnetism or diamagnetism. The electron gas has paramagnetic properties because of the electron spin and diamagnetic ones because of the orbital motion. Calculations show that the corresponding magnetic susceptibilities are $\chi_{\text{para}} \simeq \mu_B^2 n(E_F)$ and $\chi_{\text{dia}} \simeq -\frac{1}{3}\mu_B^2 n(E_F)$, where $\mu_B = e\hbar/mc$ is the Bohr magneton, i.e., the magnetic moment of an electron, and $n(E_F)$ is the density of states around the Fermi level E_F. Thus, $|\chi_{\text{dia}}/\chi_{\text{para}}| \simeq \frac{1}{3}$, so that a free-electron gas is always para-magnetic. In practice, however, there exist diamagnetic metals, such as copper, silver, and gold, the electrical properties of which are described well by the free-electron model. The qualitative explanation for this is that electrons in a crystal lattice move under the influence of the electric field of the ions. The simplest way to take account of the effect of this field on the orbital motion of the electrons is to replace the electron mass m by an effective mass m^* in the expression for the magnetic moment. On the other hand, the paramagnetic contribution to the susceptibility associated with the electron spin is not affected by external conditions. Thus, since μ is inversely proportional to the electron mass, $|\chi_{\text{dia}}/\chi_{\text{para}}| \simeq \frac{1}{3}(m/m^*)^2$. If m^* is sufficiently small, this ratio can be greater than unity, so that the metal will be diamagnetic.

Different Types of Elementary Excitations in Solids

In Section 1.1, we presented an example where the more rigorous theory (in that case, the van Hove theorem) contradicted the previous approximate one (the van der Waals theory). Here, we have an example of the opposite situation, where the later, more rigorous, theory supports the previous model, even though it reinterprets it in terms of elementary excitations instead of free electrons.

It turns out that the concept of elementary excitations can be applied to any quantum system of interacting particles. Small disturbances of such a system can be treated as an ideal gas of some quasi-particles, namely, the elementary excitations. The method of elementary excitations solves not only the problem of finding the energy of a system, but also the statistical problem of the dependence on temperature of its properties, since the statistical properties of the ideal gas are well known. We now present a few examples of quasi-particles.

(1) Spin waves, or in quantum-mechanical language "ferromagnons," are the quasi-particles that describe the departure from saturation of the magnetic moment of a ferromagnetic material. These quasi-particles, although they determine the properties of a system of interacting electrons, are described by Bose–Einstein statistics, and not by the Fermi–Dirac statistics that describe the behavior of electrons.

(2) Holes in semiconductors and insulators describe the properties of materials with not quite full valence bands of electrons. In contrast to the electrons, these quasi-particles have a positive charge.

(3) Polarons are electrons or holes in polar crystals that move through the crystal together with a polarization of the surrounding ionic lattice that they themselves generate. The effective mass of such a quasi-particle can be several hundred times the mass of a free electron.

(4) The existence of elementary excitations in liquid helium enables us to explain its superfluidity. These quasi-particles are characterized by a linear dependence of the energy on the wave vector, in contrast to the quadratic one of a free particle.

(5) Paired electrons in superconductors appear as a result of an effective attraction between electrons, induced by phonon exchange greater than their mutual Coulomb repulsion. The associated quasi-particles have an energy spectrum with a gap in it.

In all these cases, the system of interacting particles can be described by an ideal gas of some quasi-particles, the properties of which may be very different from those of the original particles.

The elementary excitations that we have described are not just mathematical tricks to permit a simple description of complicated systems. They can be observed experimentally, and the energy and momentum associated with a single excitation can be deduced, by means of probes that interact with the system so as to create or destroy individual excitations. For the elementary excitations in solids and liquids, one of the most important such probes is thermal neutrons, which have energies and momenta of the same order of magnitude as those of the excitations and a simple relationship between their energy and momentum. Another important probe, especially for the energies of the excitations, consists of electromagnetic radiation if this is absorbed or emitted by the system, while in some cases sound waves and ultrasonic waves can also be useful. These probes generally show that the excitations do not have a single sharp energy, but rather a distribution of energies of finite width. According to Heisenberg's uncertainty principle, such a spread of energies is associated with a finite lifetime of the corresponding excitation. This finite lifetime is due to the small nondiagonal elements $\gamma H'_{\alpha\beta}$ of Eq. (12), which arise from the higher-order terms in the Taylor series of Eq. (4) and which we neglected. It is only to a first approximation that the elementary excitations can be regarded as independent and so behave like an ideal gas.

1.5 Steady-State Space-Charge-Limited Currents in Insulators

It is not only for problems treated by the microscopic approach that the construction of appropriate models is important. Even for systems in which the microscopic effects are treated in terms of average, macroscopic properties of the system, such as electrical and thermal conductivities, elastic constants, and viscosity, models are frequently needed in order to analyze and understand the behavior of the system. As an example of such a situation, we consider in this section a problem in applied solid state physics, namely, the steady-state space-charge-limited currents in insulators containing traps. Our treatment emphasizes the role of models in the solution of the problem rather than technical details.

Description of the System

We consider, for convenience, a block of insulating material of rectangular cross section in the yz plane, and extending from $x = 0$ to $x = L$ in the x-direction. The two ends of the block, $x = 0$ and $x = L$, are covered with electrodes of the same metal, and a battery or some other source of emf is connected to them so that the one at $x = 0$ is the cathode, i.e., is at a negative potential relative to the one at $x = L$, the anode. The electric field associated with this potential difference will have two effects. First, it will accelerate any free electrons present in thermal equilibrium in the insulator and so lead to a steady current proportional to the applied voltage, just as for metals or semiconductors. Second, it will draw electrons from the cathode into the insulator thereby creating a region of charge imbalance in the insulator, near the cathode, known as the space-charge region. Some of this injected charge will consist of free electrons, and these will be accelerated by the field

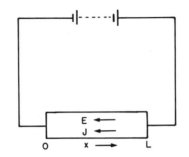

Fig. 3.   Circuit diagram for space-charge-limited currents.

just as are the ones present in thermal equilibrium. The rest of the injected charge will consist of electrons situated in trapping levels. Both these types of space-charge electrons combine to produce an electric field that tends to repel other electrons and so reduces the electric field at the cathode. For simplicity, we consider a system with electrodes such that only electrons are injected into the insulator (i.e., one carrier injection). A diagram of the system is shown in Fig. 3.

Construction and Analysis of an Idealized Model

In order to calculate the current–voltage characteristics of this system, we must first make a number of simplifying assumptions and approximations, or in other words replace the real system by a model one. Our first assumption is that the insulator has a large enough cross section for edge effects to be negligible, so that the charge densities, electrostatic potential, and electric field are all functions only of x. Thus, our model system is essentially one dimensional, and so we will have to treat only ordinary differential equations and not the much more difficult problem of partial ones. Second, we assume that the free electrons in the insulator have a constant mobility μ† independent of their energy and of the applied field, and so that their diffusion constant D is also a constant. Such an assumption is generally believed to be true for sufficiently low applied fields. Finally, we assume that the trapped electrons are completely immobile, that all have the same energy, which is E_t less than the minimum energy of free electrons, and that in the steady state the trapped and free electrons are in thermal equilibrium.

Let us denote by J the current density and by E the electric field, in the negative x-direction, as shown in Fig. 3. The current consist of two parts,

† The mobility of a charge carrier is the ratio of the magnitude of its mean velocity in the direction of an electric field, a quantity known as the drift velocity, to the magnitude of the field. The diffusion constant D is connected with this by Einstein's relation, Eq. (20).

one due to the applied field and the other to diffusion. Thus,

$$J = e\mu nE - eD \, dn/dx = \text{const}, \tag{1.14}$$

where e is the charge on an electron and $n(x)$ the density of free electrons at position x. Let $n_t(x)$ denote the density of trapped electrons there, and n_0, n_{t_0} the values of n and n_t in the bulk, neutral, insulator in thermal equilibrium. Gauss' equation is then just, in S.I. units,

$$(\varepsilon/e) \, dE/dx = n - n_0 + n_t - n_{t_0}, \tag{1.15}$$

where ε is the static dielectric constant of the insulator. Finally, the relationship between n and n_t is determined by our assumption of thermal equilibrium between them. Since both the trapped and free electrons obey Fermi–Dirac statistics, this leads to the equation

$$n_t(x) = N_t n(x)/[n(x) + N_c \exp(E_t/k_B T)] \tag{1.16}$$

where N_t is the total trap density, N_c the effective density of free electron states at temperature T, and k_B is Boltzmann's constant as usual. If we substitute in Eq. (15) for E from Eq. (14) and for $n_t(x)$ from Eq. (16) we obtain a nonlinear second-order differential equation for $n(x)$,

$$eD[(dn/dx)^2 - n \, d^2n/dx^2] + J \, dn/dx + (\mu e^2/\varepsilon)n^2(n - n_0 + n_t - n_{t_0}) = 0, \tag{1.17}$$

where n_t and n_{t_0} are given by Eq. (16). From the solution $n(x)$ of Eq. (17) we can derive $E(x)$ and J. It is not easy to write down the boundary conditions on $n(x)$ for this equation. A necessary condition is that the field $E(x)$ derived from it must satisfy

$$\int_0^L E(x) \, dx = V_0, \tag{1.18}$$

where V_0 is the potential difference between the electrodes, but this is not a very useful condition. If the insulator were infinitely long, i.e., in the limit $L \to \infty$, another condition would be

$$n(x) \to n_0 \qquad \text{as} \quad x \to L, \tag{1.19}$$

but the validity of such a condition for a finite specimen is not readily checkable.

Simplification of the Model

The above system of equations is much too complicated to solve, since we cannot find the general solution of Eq. (17) and we do not even have sufficient boundary conditions to find readily a numerical solution. To obtain a further approximation that makes the problem tractable, we examine Eq. (14). In the absence of current flow, e.g., if the circuit were open, the conduction and diffusion currents must balance and this leads to the Einstein relation

$$\mu = eD/k_B T. \tag{1.20}$$

However, if a potential is applied and the circuit closed, so that an appreciable current flows with $J > 0$, the conduction term in Eq. (14) must be much greater than the diffusion term. Thus, it is customary to ignore the diffusion term in Eq. (14) and write

$$J = e\mu n E, \tag{1.21}$$

an approximation that is justified if

$$eE(x)/k_B T \gg d(\ln n)/dx \tag{1.22}$$

as can be seen by substituting from Eq. (20) into Eq. (14). Such an approximation means that we are now considering a model system in which there is no diffusion of electrons, but only their field-induced motion. Thus, we set $D = 0$ in Eq. (17), which thereby becomes a first-order equation. A single boundary condition is required to solve this, and the one normally used is to specify the electric field at the cathode,

$$E(0) = E_c. \tag{1.23}$$

Frequently, E_c is set equal to zero. In such a case, $n(0)$ is infinite according to Eq. (21), so that near the cathode the main current arises from diffusion, and condition (22) is certainly not satisfied. In spite of this, it is found that the model system represented by Eq. (17) with $D = 0$ and Eq. (23) does have current–voltage characteristics similar to those of the real system. We return to this approximation in Section 5.4.

Solutions for Extreme Cases

Even this simplified model cannot readily be solved, but an examination of extreme cases can provide us with a lot of information about the solution. Thus, if we consider a model system with no injected charge, $n = n_0$ throughout, Eq. (21) gives us $J = e\mu n_0 E$. Since injected charge can only increase the current, we see that

$$J > e\mu n_0 E. \tag{1.24}$$

Another extreme case is that in which the injected free carriers are so much more numerous than those present in the neutral insulator that n_0 can be neglected. If, in addition, there are no traps, i.e., $N_t = 0$, one readily finds that

$$J \leq 9\varepsilon\mu V_0^2/8L^3, \tag{1.25}$$

where the equality occurs only for $E_c = 0$. Since the presence of traps can only reduce the density of injected free charge, and hence the current, we see that inequality (25) must hold in all cases. Thus, these two extreme cases, or model systems, provide us with lower and upper bounds on the current for a given V_0 in any real situation. Another extreme case from which valuable information can be obtained is the trap-filled case, a hypothetical system in which all the N_t traps are filled with injected charge before any free electrons are injected into the solid. For this system, of course, Eq. (16) is replaced by $n_t(x) = N_t$, and the resulting system of equations can be solved exactly. It provides another lower bound for J as a function of V_0, since in practice some free electrons must be injected along with the trapped ones. Further information, and not just bounds on J, can also be obtained from these limiting cases. For instance, if V_0 is sufficiently small, the amount of injected charge must be low, so that Eq. (24) will be a good approximation to the solution. On the other hand, if V_0 is high enough for $n(x)$ to be much greater than $N_c \exp(E_t/k_B T)$, Eq. (16) shows us that $n_t(x) \approx n_t$, i.e., we have the trap-filled case. Similarly, if V_0 is large enough for $n(x) \gg N_t$, the trapped charge will play only a minor role and the situation will be essentially that of the trap-free system. Such use of models to obtain bounds on the behavior of a system and qualitative ideas of how it behaves is quite common and extremely valuable.

1.6 Boundary Layer Theory in Hydrodynamics

We now turn to another example of a macroscopic problem for which the use of the appropriate model is vital, namely, the flow of a fluid past a solid surface. The hydrodynamic equations describing the motion of a fluid are nonlinear, as was the system of equations considered in the last section, but here the method of constructing a model is somewhat different and its use is very different.

The Equations of Motion for a Fluid

For an incompressible fluid with a constant viscosity coefficient η, the velocity $\mathbf{v}$ of an element of fluid must satisfy the Navier–Stokes equation

$$\partial\mathbf{v}/\partial t + (\mathbf{v} \cdot \nabla)\mathbf{v} = -\nabla p/\rho + (\eta/\rho) \nabla^2\mathbf{v}, \tag{1.26}$$

where p is the pressure and ρ the density. For an incompressible fluid, the continuity equation is just $\partial\rho/\partial t = 0$, and so

$$\nabla \cdot \mathbf{v} = 0. \tag{1.27}$$

A problem is completely defined by this set of equations together with the appropriate boundary conditions. At a solid surface, one such condition is that the fluid cannot penetrate the surface: if $\mathbf{n}$ denotes the normal to the surface, this condition can be written as

$$\mathbf{v} \cdot \mathbf{n} = 0. \tag{1.28}$$

In addition, there are attractive intermolecular forces between a viscous fluid and a solid surface, and these will cause the tangential component of the velocity to vanish at the surface, i.e.,

$$\mathbf{v} \times \mathbf{n} = 0 \tag{1.29}$$

Sometimes, the viscosity of a fluid can be neglected, in which case the fluid is called ideal. This is only possible if the viscosity term $(\eta/\rho)\,\nabla^2\mathbf{v}$ in Eq. (26) is very much smaller than the inertia term $(\mathbf{v} \cdot \nabla)\mathbf{v}$. If U and l are some characteristic velocity and length for a given problem, this inequality can be written as $U^2/l \gg (\eta/\rho)U/l^2$, i.e., the Reynold's number $\mathrm{Re} = Ul/(\eta/\rho) \gg 1$. For an ideal incompressible fluid, Eq. (26) becomes

$$\partial\mathbf{v}/\partial t + (\mathbf{v} \cdot \nabla)\mathbf{v} = -\nabla p/\rho, \tag{1.30}$$

while Eq. (27) is unaltered. For this set of equations, only boundary condition (28) applies, and not (29) as well, since in the absence of viscous forces we cannot make any statement about the tangential velocity of the fluid at the surface. In fact, since Eq. (30) is only a first-order partial differential equation while Eq. (26) was of second order, the solutions of the equation for the ideal fluid cannot be required to satisfy the additional boundary condition (29), and in general will not do so.† However, neglect of the viscous forces means that the theory of an ideal fluid cannot explain the resistive force acting on a body moving uniformly through a fluid (the D'Alambert paradox).

The nonlinearity of Eq. (26) makes it difficult to find an exact solution. Such solutions can be found mainly for systems in which the nonlinear term vanishes because of the symmetry (Poiseuille flow, Couette flow, etc.), and for a few other special cases, such as fluid flow near a rotating disk (see Problem 2.9) and the submerged jet. In most cases, however, only approximate solutions can be found, and for these it is essential to use a suitable model of the system.

---

† We shall return in Section 5.4 to an analysis of systems of equations in which a small parameter multiplies the highest-order derivative in a differential equation.

The Flow of Fluid past a Solid Body

A typical problem, which we shall now consider, is that of the steady motion of a solid body through a fluid, or equivalently the flow of fluid past a solid surface, for a fluid with a Reynold's number much greater than unity. For such a fluid, the viscosity is very small, so that we might expect the model of an ideal fluid to be appropriate. However, as we have seen, such a model does not enable us to satisfy boundary condition (29) at the surface of the solid. On the other hand, Eq. (26) is so much more complicated than Eq. (30) that we most certainly do not want to have to solve it when considering a fluid of very low viscosity. The appropriate compromise that produces a solvable model was found by Prandtl with the theory of the boundary layer. According to this model, the motion of the fluid is divided into two regions. The motion of the main part of the fluid can be described by Eq. (30) for an ideal fluid, as if there were no solid boundary. However, in a narrow region near the surface which is known as the boundary layer, Eq. (26) has to be used. For this region, because of its small thickness, the equation can be simplified into a solvable form. In this way, it is possible to satisfy the boundary condition of Eq. (29) and also provide an analytical continuation of the solution in the boundary layer region to the region of ideal fluid flow.

As an example of this process, we now examine the simplest possible case, namely, the steady flow of such a fluid over a flat plate. We choose the x axis in the direction of flow and the origin so that the plate occupies the plane $y = 0$, $x \geq 0$, as shown in Fig. 4. For the sake of simplicity, we assume that the system is essentially infinite in the z-direction, so that the velocity of the fluid is independent of z. The continuity and Navier–Stokes equations then take the form:

$$\partial v_x/\partial x + \partial v_y/\partial y = 0, \tag{1.31}$$

$$v_x\,\partial v_x/\partial x + v_y\,\partial v_x/\partial y = (-1/\rho)\,\partial p/\partial x + (\eta/\rho)(\partial^2 v_x/\partial x^2 + \partial^2 v_x/\partial y^2), \tag{1.32a}$$

$$v_x\,\partial v_y/\partial x + v_y\,\partial v_y/\partial y = (-1/\rho)\,\partial p/\partial y + (\eta/\rho)(\partial^2 v_y/\partial x^2 + \partial^2 v_y/\partial y^2). \tag{1.32b}$$

Simplification of the Hydrodynamic Equations

It is possible to simplify this set of equations, to a good approximation, by taking account of the small thickness δ of the boundary layer. This is much less than the characteristic length l of the problem under consideration, which for the real system can be taken as the length l of the plate. Hence, in

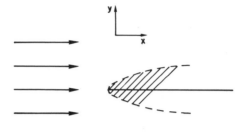

Fig. 4.   Flow of fluid past a flat plate; the hatched region denotes the boundary layer.

the boundary layer $\partial/\partial x$ will generally be of order $1/l$, and so much smaller than $\partial/\partial y$, which will be of order $1/\delta$. As a result, Eq. (1.32a) can be replaced by the much simpler boundary layer equation

$$v_x\,\partial v_x/\partial x + v_y\,\partial v_x/\partial y = (\eta/\rho)\,\partial^2 v_x/\partial y^2, \tag{1.33}$$

which has to be solved in conjunction with Eq. (31). The boundary conditions for this system of equations are that at the surface of the plate

$$v_x = v_y = 0, \qquad \text{for} \quad y = 0 \quad \text{and} \quad x \geq 0, \tag{1.34}$$

while at large distances from the plate the velocity must be that in the main stream of the fluid, so that

$$\begin{aligned} v_x &\to U & \text{as} \quad y \to \pm\infty, \\ v_y &\to 0 & \text{as} \quad y \to \pm\infty. \end{aligned} \tag{1.35}$$

In order to solve this set of equations, we first define a stream function $\phi(x, y)$ by means of

$$v_x = -\partial\phi/\partial y, \qquad v_y = \partial\phi/\partial x, \tag{1.36}$$

so as to solve the continuity equation (31). We then write

$$w = y(x\eta/U\rho)^{-1/2}, \tag{1.37}$$

a change of variables of a form discussed in the next chapter, and define a function $f(w)$ by

$$\phi(x, y) = (\eta U x/\rho)^{1/2} f(w). \tag{1.38}$$

Simple calculations then show that $f(w)$ satisfies the ordinary differential equation (Blausius' equation)

$$2f''' + ff'' = 0 \tag{1.39}$$

with the boundary conditions

$$f(0) = f'(0) = 0, \qquad f'(\pm\infty) = 1. \tag{1.40}$$

This equation can be solved numerically, and this solves the problem of boundary layer flow past a flat plate in general. The properties of any specific system enter the solution only through the scaling factors in Eqs. (37) and (38).

The boundary layers for the laminar and turbulent flow of fluids play an important role in the motion of airplanes, ships, and cars, the working of pumps and turbines, and numerous other systems. The solution of all such problems is made possible by the use of a suitable model, in which the narrow boundary layer (in which the flow is nearly parallel to the surface and varies rapidly as one moves away from the surface) is treated differently from the main stream.

This use of two different models for two different regions of space in the same system is very different from the use of different models for the whole system according to the value of an external parameter V_0 that was considered in the last section. The common feature in both cases is that a model system is found which is similar to the real one, but obeys a much simpler set of equations which can be solved much more readily than those for the original system.

Chapter 2 | Dimensional Analysis

2.1 Introduction

Dimensional analysis is based on two simple features of physical formulas. First, such formulas are not just mathematical equations that relate numbers; they are equations that relate physical quantities. Every physical quantity can be described by the product of a number and a unit, and these units have dimensions. Second, physical formulas and equations must be dimensionally homogeneous, i.e., every term must be of the same dimensions. While these statements are fairly obvious, it is surprising how much information can be derived from the dimensional analysis of problems.

Fundamental and Derived Units

An essential preliminary step before any analysis of dimensions is to assign the appropriate dimensions to all physical quantities. We denote the dimension of a quantity by enclosing the symbol for it in square brackets, i.e., $[x]$ denotes the dimensions of the quantity x with no regard to its numerical value. Physical quantities are generally divided into two groups, according to whether their dimensions are fundamental (or primary, or basic) ones or derived (or secondary) ones. The dimensions of the derived quantities are expressed in terms of the dimensions of the fundamental quantities with the aid of the appropriate physical formulas in which the dimensions of all quantities except the one under consideration are known. The division into fundamental and derived quantities is to some extent arbitrary and a matter of convenience. For instance, while mass M, length

L, and time T are generally treated as fundamental units, quantities such as force F and temperature θ can be chosen to be fundamental or derived ones. An examination of these two examples will clarify the advantages and disadvantages of increasing the number of fundamental dimensions.

Since acceleration a is of the form d^2x/dt^2, its dimensions are LT^{-2}. Thus if we write Newton's second law in the form

$$F = kma, \tag{2.1a}$$

we see, on taking dimensions, that

$$[F] = [k]MLT^{-2}. \tag{2.1b}$$

If we wish, we can choose force to be a derived quantity and k to be dimensionless, in which case

$$[F] = MLT^{-2}. \tag{2.2}$$

Alternatively, we could choose to define an extra *basic* dimension $[F]$, in which case k would be a dimensional factor with dimensions $[F](MLT^{-2})^{-1}$. A similar situation exists with regard to temperature τ, which is related to mechanical quantities according to the theorem of the equipartition of energy

$$\tfrac{1}{2}k_B\tau = \tfrac{1}{2}m\overline{v^2} \tag{2.3a}$$

so that

$$[k_B][\tau] = ML^2T^{-2}. \tag{2.3b}$$

Here, too, we can choose k_B to be dimensionless, in which case the temperature τ is of dimension ML^2T^{-2}, just like energy, or we can define a new basic dimension θ, in which case k_B has dimensions $ML^2T^{-2}\theta^{-1}$. From these two examples, we see that the greater the number of fundamental dimensions, the greater the number of dimensional factors in physical equations. Such extra factors complicate the equations and so are undesirable. On the other hand, the smaller the number of fundamental dimensions, the greater is the number of derived quantities that have the same dimensions, e.g., temperature and energy if we choose k_B to be dimensionless. This is also undesirable, since it blurs the distinction between quantities that have a different significance (thereby also reducing the usefulness of dimensional analysis) and can impose an undesirable system of units on the extra derived quantities. Thus, some sort of compromise has to be found. For the mechanical and thermal properties of systems, it is found most convenient to use just the four basic dimensions M, L, T, and θ. As a result, Boltzmann's constant k_B does have dimensions, but Newton's second law is just $F = ma$.

For problems involving electric and magnetic quantities, there are several different choices of dimensions in common use. The electrostatic system is based on Coulomb's law for the force F between charges q_1 and q_2, distance r apart, in a medium of dielectric constant ε,

$$F = q_1 q_2/\varepsilon r^2. \tag{2.4}$$

If ε is chosen to be dimensionless, then the dimension of electric charge q is $M^{1/2}L^{3/2}T^{-1}$, so that the dimension of current $I \sim dq/dt$ is $M^{1/2}L^{3/2}T^{-2}$. The electromagnetic system, on the other hand, is based on the Biot–Savart law for the force F between two current elements $I\,dl$ and $I'\,dl'$, distance r apart, in a medium of magnetic permeability μ,

$$F = \mu II'\,dl\,dl'/r^2. \tag{2.5}$$

If μ is chosen to be dimensionless, the dimension of current is just $M^{1/2}L^{1/2}T^{-1}$. Thus, while either the electrostatic or the electromagnetic system is self-consistent, they are not consistent with each other unless we ascribe dimensions to ε and/or μ. If we do this, the dimensions of current in the two systems are, respectively, $[\varepsilon]^{1/2}M^{1/2}L^{3/2}T^{-2}$ and $[\mu]^{-1/2}M^{1/2}L^{1/2}T^{-1}$. These will be equal if $[\varepsilon\mu] = (LT^{-1})^{-2}$, i.e., if $\varepsilon\mu$ has the dimensions of $[\text{velocity}]^{-2}$. Hence, if we denote their values in vacuum by the subscript 0, we can write $\varepsilon_0\mu_0 = 1/c^2$, where c is a velocity; in fact, c is just the velocity of light in vacuum. While the dimension of either ε or μ could be chosen as an extra basic dimension, either choice would lead to charge and current having dimensions with fractional exponents. In order to avoid this, the internationally recommended S.I. system of units chooses the current I as an extra fundamental dimension.

Finally, for problems involving the intensity of illumination, an additional fundamental unit is introduced, namely, the intensity Q of a source of light. Thus, dimensional analysis is generally based on the six fundamental dimensions M, L, T, θ, I, and Q.

Derivation of Formulas

As we mentioned at the beginning of this chapter, a necessary (but not sufficient!) condition for the accuracy of a physical equation is that all the terms in it shall be of the same dimension. This fact not only enables us to check given equations but also in some cases allows us to derive them, or at least their form. The simplest example is that of calculating the terminal velocity v of a body of mass m falling freely to the ground from height h in a gravitational field that produces acceleration g per unit mass. Since v can depend only on the above-mentioned properties of the system, we write

$$v = f(h, g, m). \tag{2.6}$$

While we do not know the form of f, we can postulate that it is the product of powers of h, g, and m, so that

$$v \sim h^{\alpha} g^{\beta} m^{\gamma}. \tag{2.7a}$$

On taking dimensions of this equation, we find that

$$LT^{-1} = L^{\alpha}(LT^{-2})^{\beta}M^{\gamma}, \tag{2.7b}$$

so that on equating powers of L, T, M, we obtain

$$\alpha = \beta = \tfrac{1}{2}, \qquad \gamma = 0, \tag{2.8}$$

i.e.,

$$v \propto \sqrt{gh}. \tag{2.9}$$

Thus, the terminal velocity is independent of the mass of the body. On the other hand, the gravitational acceleration g is important, and the problem could not have been solved without taking it into account. In a similar way, the velocity of light c must be taken into account in the dimensional analysis of electromagnetic problems.

The above example shows us two basic limitations of the method of dimensional analysis. First, it can only lead to the relationship between the powers of different quantities if such a functional dependence is postulated. Incidentally, more complicated forms of dependence can sometimes be excluded because of the analytic properties of the problem, as we discuss in Chapter 4. Second, dimensional analysis allows us at best to find a physical formula or equation up to a constant of proportionality, but we cannot find these constants since they are dimensionless. Such constants are frequently found to be of order unity, as in the above problem where the exact solution is $v = (2gh)^{1/2}$, but there are some exceptions.

A similar problem of dimensionless quantities arises when among the variables characterizing a system there are several having the same dimension. For instance, in the problem of the motion of a planet of mass m_1 around the sum of mass m_2, which we discuss in Section 2.2, the ratio m_1/m_2 is a dimensionless quantity and so beyond the scope of dimensional analysis. Thus, in this case an arbitrary function of m_1/m_2 must be inserted in the results obtained from dimensional analysis, and some other methods used to find this function.

Dimensionless parameters can also appear as a dimensionless combination of some characteristic variables of the problem. Let us consider, for instance, the problem of the force acting on a body moving through a viscous incompressible fluid with the constant velocity U. We assume that the body has a characteristic size l, an approximation which is exact for a sphere of radius l. The resistive force F can depend only on l, U, and the density ρ and viscosity η of the fluid, so that

$$F = F(l, U, \rho, \eta). \qquad (2.10)$$

The number of arguments of the above function is four, one more than the number of fundamental mechanical dimensions, in contrast to our previous example. On equating the powers of the dimensional equation corresponding to Eq. (10), we obtain only three equations relating the powers of the four arguments in this equation. Hence, the solution will contain an arbitrary function of some dimensionless parameter which is a combination of these variables. In order to derive this parameter, we note that the viscosity of a fluid is defined in terms of the resistive force per unit area ε, as a result of a velocity gradient ∇v,

$$\varepsilon = -\eta \, \nabla v, \qquad (2.11a)$$

so that

$$[\varepsilon] = [\eta][v]L^{-1}, \qquad (2.11b)$$

i.e.,

$$[\eta] = (ML^{-1}T^{-2})(LT^{-1})^{-1}L = ML^{-1}T^{-1}. \qquad (2.11c)$$

We now look for a dimensionless parameter Re as a function of all our variables, and so write

$$\mathrm{Re} = U^\alpha l^\beta \rho^\gamma \eta^\delta. \qquad (2.12)$$

We readily find that Re, called the Reynolds number, is of the form

$$\mathrm{Re} = Ul\rho/\eta \qquad (2.13)$$

or some power of this combination of parameters. Thus, the dimensionless parameter Re will enter the final result as an arbitrary parameter, just as m_1/m_2 did in the case of planetary motion.

Numerous different examples of the use of dimensional analysis to derive physical formulas will be presented in Section 2.2.

Nonlinear Heat Conduction

So far, we have only considered how physical formulas can be derived by dimensional analysis. However, this analysis, and in particular the use of dimensionless variables, often provides us with an insight into the physical relationships of the quantities appearing in a problem which helps us to solve it. In many cases, the individual parameters are not important, but only certain combinations of them. The use of these combinations to define new variables can reduce the number of arguments and independent variables in a problem, and thereby simplify its solution.

A nontrivial example of this procedure is provided by the analysis of the nonlinear heat conductivity equation. The general form of the heat conductivity equation for one-dimensional problems is

$$C\, \partial\theta/\partial t = (\partial/\partial x)(K\, \partial\theta/\partial x), \tag{2.14}$$

where $\theta(x, t)$ denotes the temperature at point x at time t. If the specific heat C is constant and the thermal conductivity K is proportional to θ^n, Eq. (14) becomes of the form

$$\partial\theta/\partial t = (\partial/\partial x)(a\theta^n\, \partial\theta/\partial x), \quad \text{where} \quad K/C = a\theta^n. \tag{2.15}$$

If the specific heat C is also a function of temperature, an equation of this form can still be derived by means of the change of variable $\theta' = \int C(\theta)\, d\theta$.

We look for a solution of Eq. (15) of the form

$$\theta(x, t) = At^{-k}f(w), \quad w = x/Bt^m, \tag{2.16}$$

where A and B are constants, w is a dimensionless variable, and $f(w)$ is an arbitrary function of w. If we substitute this solution in Eq. (15) and take dimensions of the resulting equation, we readily find that

$$t^{-(k+1)} \propto aA^n B^{-2} t^{-(2m+nk+k)}. \tag{2.17}$$

Since a, A, and B are all independent of time, on equating dimensions we find that

$$2m + nk - 1 = 0, \tag{2.18}$$

and that $aA^n B^{-2}$ is dimensionless. Since the magnitude of A and B is undefined, we can choose it so that

$$aA^n B^{-2} = 1. \tag{2.19}$$

The set of equations (15)–(19) readily leads to an equation for $f(w)$,

$$(d/dw)(f^n\, df/dw) + mw\, df/dw + kf = 0, \tag{2.20}$$

which depends only on the values of $k, m,$ and n. While n is known for a given problem, Eq. (18) gives only one relationship between k and m, so that another equation is required to determine them uniquely.

The missing relationship between k and m can be found from the boundary conditions. The total thermal energy of the system, E, per unit cross section is given by

$$E \propto \int_{-\infty}^{\infty} \theta\, dx = ABt^{m-k} \int_{-\infty}^{\infty} f(w)\, dw. \tag{2.21}$$

Various situations can arise, in accordance with the boundary conditions. For instance, if E is constant, which corresponds to the diffusion of a fixed quantity of thermal energy, we find, using Eq. (18) also, that

$$m = k = 1/(n + 2). \tag{2.22}$$

Alternatively, we can consider a system into which heat is introduced at a constant rate, for instance, by chemical or radioactive processes. In this case, $E \propto t$, i.e., $m - k = 1$, so that

$$m = (n + 1)/(n + 2), \qquad k = -1/(n + 2). \tag{2.23}$$

Another possibility is that the temperature at the point $x = 0$ is held constant, so that $\theta(0, t) = \theta_0$. As we can see from Eq. (16), in such a case,

$$k = 0, \qquad m = \tfrac{1}{2}. \tag{2.24}$$

The problem is finally solved by substituting the appropriate values of k, m, and n into Eqs. (16) and (20).

In all cases, the solutions are of the thermal wave type in which the disturbance propagates with a finite velocity if n differs from zero. For linear heat transfer processes, on the other hand, i.e., if $n = 0$, the solution of Eqs. (16), (20), and (22) is of the form $\theta \sim t^{-1/2} \exp(-x^2/4at)$, i.e., temperature changes on the boundary produce an instantaneous response at every point of the system. This instant response corresponds to an infinite velocity of propagation of the disturbance and is to be contrasted with the finite velocity and consequent delayed response in the nonlinear case. We have derived this result by a successful choice of dimensionless variables that enabled us to reduce the nonlinear partial differential Eq. (15) to the much simpler nonlinear ordinary differential equation (20). This is just the procedure that we used in reducing the boundary layer equation to the Blausius equation in Section 1.6, and it can also be applied in many other problems.

In Section 2.3 we consider two types of examples. First, we use a change of scale to derive the form of the laws of motion for particles moving in different potentials. After this, we use dimensionless variables in statistical physics to derive equations of state for various systems.

Dimensionless Equations

When we discussed boundary layer theory in Section 1.6, we saw that the differential equation (1.39), together with its boundary conditions (1.40), do not contain any parameters relating to a given system. Hence, its unique numerical solution can be applied to any system by means of an appropriate change of variables. The problem of nonlinear heat conduction that we have just considered is also of this type, since Eq. (20), for given values of k, m, and n, is applicable to all systems of the same sort. Another example of the value of dimensionless variables is provided by the van der Waals equation, which we considered in Section 1.1. This equation,

$$(p + a/V^2)(V - b) = RT, \qquad (2.25)$$

contains parameters a and b that vary from one fluid to another. Because of these parameters, it can provide a qualitative description of phase transitions, and in particular predicts a critical point† (p_c, V_c, T_c), which is characterized by the conditions

$$(\partial p/\partial V)_{T_c} = (\partial^2 p/\partial V^2)_{T_c} = 0. \qquad (2.26)$$

One readily find that these critical parameters are given in terms of a and b by the formulas

$$V_c = 3b, \qquad p_c = a/27b^2, \qquad RT_c = 8a/27b. \qquad (2.27)$$

If we now introduce dimensionless reduced parameters

$$\pi = p/p_c, \qquad \tau = T/T_c, \qquad \text{and} \qquad \phi = V/V_c, \qquad (2.28)$$

the van der Waals equation becomes

$$(\pi + 3/\phi^2)(3\phi - 1) = 8\tau. \qquad (2.29)$$

Equation (29), in contrast to Eq. (25), contains no parameters that depend on the specific gas considered, and so is equally valid for all gases.

The existence of an equation such as (29) leads immediately to the law of corresponding states. According to this law, if a pair of substances have the same values of two of the reduced variables, then the value of the third variable must also coincide. Thus, the reduced pressure is a unique function of the reduced volume and temperature, $\pi = \pi(\tau, \phi)$. The law of corresponding states applies not only to the van der Waals equation, but also to any other equation of state that involves not more than three parameters. Thus, this law is somewhat more accurate than the van der Waals equation, although its range of applicability is also restricted.

There are many other examples of the use of dimensionless variables. For instance, the Thomas–Fermi equation for the electronic charge distribution in a heavy atom can be put into a dimensionless form, and the numerical solution of the resulting equation used to find the electrostatic potential and charge distribution for any atom. This problem and a problem concerning heat conduction in a cube are considered in the first part of Section 2.4.

---

† The critical point for a system having two phases is the one beyond which these phases are indistinguishable.

Hydrodynamic Modeling

In the examples discussed so far, dimensional analysis led to equations with constant coefficients, such as Blausius' equation (1.39) or the reduced van der Waals equation (2.29), and these equations can then be solved numerically. However, in other cases, the use of dimensionless variables can lead to equations that contain, in addition to numbers, some dimensionless combination of the initial variables. For such systems, the concept of modeling arises naturally. The most familiar example of this concept is in the field of hydrodynamics, where it serves as the basis of wind-tunnel experiments, and so we will briefly outline the ideas involved in connection with such problems.

For the steady flow of an incompressible viscous fluid, one readily finds from Eq. (1.26) that

$$-\nabla \times (\mathbf{v} \times (\nabla \times \mathbf{v})) = (\eta/\rho)\, \nabla^2(\nabla \times \mathbf{v}). \tag{2.30}$$

If U and l are a typical (constant) velocity and length for the problem, we introduce dimensionless variables

$$\mathbf{c} = \mathbf{v}/U, \qquad \mathbf{h} = \mathbf{r}/l. \tag{2.31}$$

In terms of these, and the Reynold's number Re defined in Eq. (13), Eq. (30) becomes

$$\nabla' \times (\mathbf{c} \times (\nabla' \times \mathbf{c})) = (l/\mathrm{Re})\, \nabla'^2(\nabla' \times \mathbf{c}), \tag{2.32}$$

where the primes denote derivatives with respect to the dimensionless variable $\mathbf{h}$. Since Eq. (32) contains the dimensionless parameter Re, its solution leads to

$$\mathbf{v} = U f(\mathbf{r}/l, \mathrm{Re}). \tag{2.33}$$

In other words, if two systems are characterized by the same Reynolds number, the dimensionless velocities $\mathbf{v}/U$ in them will be the same functions of the dimensionless coordinates. The motions in the two systems are then said to be similar, since from the characteristics of the flow in one system we can obtain those of the other system by the use of appropriate similarity coefficients. This concept of the similarity of physical phenomena is a generalization of that of geometrical similarity, where the lengths of corresponding lines in two similar figures can be obtained one from the other by multiplying by a similarity coefficient. The existence of physical similarity is of great practical importance and is the basis of modeling, i.e., the replacement of the problem in which we are interested by the similar one for a model of a different size.

The basic idea of this modeling, in which we look for systems that have different physical parameters but behave in the same way, is similar to that of the construction of models that we discussed at length in Chapter 1. The principle in each case is to find a system that behaves in the same way as the original one but is more convenient to treat. The difference between the two cases lies in the reason why the model system is more convenient. In Chapter 1, we used models in order to reduce the complexity of the problem, and the advantage of them is the greater simplicity of their properties and quantitative analysis. Here, on the other hand, we are usually interested in a model that is of a more convenient size, more easily modified, or less expensive than the original system.

Physical similarity analogous to that of Eq. (33) can also be applied to magnitudes characterizing the motion of the fluid that are not functions of the coordinates. For instance, it follows from Eqs. (10) and (13) that the force F acting on a sphere moving with uniform velocity through an incompressible fluid depends only on the Reynolds number Re. Thus, the dimensionless ratio of F to a magnitude with the dimensions of force must be a function only of Re. Hence, if we choose $\eta U l$ as the unit of force for the problem, we can write

$$F = \eta U l f(\text{Re}). \tag{2.34}$$

The form of the functions $f(\text{Re})$ cannot be found by dimensional analysis, but only from exact calculations or from experimental results. For small velocities, Stokes' law $F = 6\pi\eta U l$ is found, i.e., $f(\text{Re}) = 6\pi$. On the other hand, it is known from experiments that for large velocities the resistive force is independent of the viscosity, and so, since Re is inversely proportional to the viscosity, $f(\text{Re}) \sim \text{Re}$, and, in fact, $F \propto \rho U^2 l^2$.

Only one dimensionless parameter, the Reynolds number Re, appears in Eqs. (33) and (34) because only one such parameter can be constructed from the quantities, ρ, η, l, and U involved in the problem. If, however, one allows the fluid to be compressible, another new dimensionless parameter will appear, while the inclusion of gravitational forces and/or the consideration of nonstationary flow patterns will lead to additional dimensionless parameters. The different types of physical similarity and examples of modeling will be considered in the last part of Section 2.4.

Phase Transitions

The concept of similarity that we have just discussed is not restricted to problems of the same type but having different magnitudes of typical lengths, velocities, and such quantities. It can also be used as the basis for the study of more complicated phenomena which occur in many different types of system. An example of this is provided by the scaling theory of transitions from an ordered to a disordered phase. Such transitions occur in many different types of problem, such as the positioning of two types of atom on the crystalline lattice of a binary alloy, the arrangement of the electrical dipole moments in ferroelectric and antiferroelectric materials and of the elementary magnetic moments in ferromagnetic, antiferromagnetic, and ferrimagnetic materials, the ordering of the electron states in superconductors and of the helium atoms in superfluid helium, and the disappearance of spatial homogeneity at the critical points of pure and multicomponent fluids.

The appearance of the same type of features in such different physical systems (classical and quantum, liquids and solids, at low and at high temperatures) strongly indicates that the order–disorder phase transition is a general property of many-body systems, and does not depend on the nature of the system. This idea is supported by simple thermodynamic considerations concerning the entropy and the energy of a system. In general, the internal energy E has its minimum value in the order state. The entropy S of the system is associated with the amount of disorder of the arrangement of the particles in the system and increases as their distribution becomes more random. For a system restricted to a fixed volume, the stable state at a given temperature T is that which minimizes the free energy

$$F = E - TS. \tag{2.35}$$

At high temperatures, the negative second term in F dominates, so that the minimum value of F is associated with the maximum entropy and so with a disordered state. At low temperatures, on the other hand, the internal energy E is the dominant term in F, so that the stable state is that with minimum E, i.e., the ordered one. The temperature of the order–disorder transition is determined by the balance between the ordering or energetic tendency and the disordering or entropic one. Thus, the appearance of order is caused by the interactions between particles (as a result of which the ordered state has the minimum internal energy), but not with the specific form of this interaction.† Hence, the order–disorder transition for different systems should have a common nature determined by the statistical properties of many-body systems.

From the theoretical physicist's point of view, the appearance of ordering means the existence of a statistical connection between particles located a long distance from each other. An interaction between neighboring par-

---

† The ideal gas has no phase transition, as stated in Section 1.1, because there are no interactions between the particles that could make the internal energy depend on their arrangement.

ticles may be responsible for this connection, since the information about the state of a given particle is transferred from one particle to another, as a result of this interaction, over long distances of the order of the correlation length ξ. For instance, suppose that in a ferromagnetic material one forces a particular spin (i.e., elementary magnetic moment) to point in some direction. The ferromagnetic interaction tries to align adjacent spins in the same direction, and in the fully ordered state the correlation length ξ will be infinite, but in a disordered state ξ is finite. As the temperature approaches that of the order–disorder phase transition, ξ must become very large, and it is then natural to expect that the details of the interactions over small distances become unimportant. If this is so, the simplest model that exhibits such a phase transition can be used to obtain a description of the system.

The Ising Model

Such a simple model is provided by the Ising model, as depicted in Fig. 1, which is based on the following three assumptions:

(1)   The objects to be ordered are located on the sites of some crystalline lattice with lattice parameter a.

(2)   Each object can be in only one of two possible different states, which are characterized by the values ± 1 for the variable s_i associated with site i (e.g., direction of spins in a ferromagnet, location of A or B atoms in a binary allow AB, etc.).

(3)   Account is taken only of interactions between nearest neighbors, a given pair (i, j) of which contribute $-Js_i s_j$ to E where J is positive, and of an external "magnetic" field H that tends to make s_i have the same sign as itself, so that

$$E = -J \sum_{(i,j)} s_i s_j - H \sum_i s_i. \qquad (2.36)$$

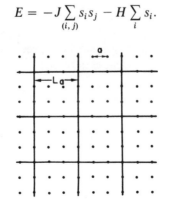

Fig. 1.   Point and "block" square lattices for the two-dimensional Ising model.

The two-dimensional Ising model was solved exactly for $H = 0$ by Onsager in 1944. He found that for $T < T_c$, the system is ordered, in the sense that $\langle s_i \rangle \neq 0$ so that more objects are in one state than in the other, while for $T > T_c$ ordering is absent. The temperature T_c of the phase transition is determined completely by the interaction energy J. However, this exact solution has been found only for two-dimensional lattices, while we are interested mainly in three-dimensional systems.

Scaling Theory

We shall now show how it is possible to use physical similarity in order to find connections between the behavior of different thermodynamical quantities, such as susceptibilities, near and at the order–disorder transition temperature T_c. These connections are known as scaling laws. In order to derive them we imagine the Ising lattice to be divided into blocks, in such a way that each block contains a large number of spins but its size remains much smaller than the correlation length, as shown in Fig. 1. These blocks will be arranged on a new lattice, the "block" lattice, with a lattice parameter La such that

$$1 \ll L \ll \xi/a. \tag{2.37}$$

Since the size of the block is much less than ξ, all the spins inside a given block will be aligned. We can therefore characterize the blocks as a whole by a variable μ_A corresponding to their total spin, and such that the energy of the block lattice is

$$E' = -J' \sum_{A,B} \mu_A \mu_B - H' \sum_A \mu_A. \tag{2.38}$$

A phase transition is described by two variables, which can be conveniently chosen as a dimensionless temperature $\tau = (T - T_c)/T_c$ and a dimensionless external field h. The only possible difference between the original site problem and the new block one is in the values of these two parameters because of the differences between (J, H) and (J', H'). Let the values of these parameters for the block problem be denoted by $\tilde{\tau}$ and $\tilde{h}$, which in general depend on the size L of the block.

In order to find the connection between the variables for the site and block problems, we note that they represent different descriptions of the same lattice, so that ordering takes place simultaneously in the two cases, while the external fields also vanish simultaneously. Hence, we can write

$$\tilde{\tau} = \tau L^x, \qquad \tilde{h} = h L^y, \tag{2.39}$$

where x and y are constants. While the solution of the problem for the block lattice is no simpler than for the original site lattice, a *comparison* of these two problems leads to some important results.

Let $f(\tau, h)$ be the singular part of the free energy per spin. Then for a system of dimensionality d, $L^d f(\tau, h)$ is the singular contribution to the free energy from a set of L^d spins of the site lattice. On the other hand, the same set of spins serves as a block in the block lattice, where this set is described by the free energy $f(\tilde{\tau}, \tilde{h})$. According to our basic hypothesis, these two systems are described by the similar Hamiltonians of Eqs. (36) and (38). Therefore the thermodynamic functions describing these two models must also be similar, provided that each of them is described by its own variables. These are just τ, h and $\tau L^x, hL^y$, respectively, in view of Eq. (39). Thus we require that

$$f(\tau L^x, hL^y) = L^d f(\tau, h). \tag{2.40}$$

Equation (40) has to be satisfied identically for all values of L obeying condition (37). This is possible only if the function $f(\tau, h)$ is of the form

$$f(\tau, h) = \tau^{d/x} \psi(\tau/h^{x/y}), \tag{2.41}$$

where ψ is a homogeneous function of τ and $h^{x/y}$. Although the explicit form of the function ψ is unknown, a number of important results can be derived solely from its homogeneity. These results concern the behavior of different physical observables near the critical point.

First, let us assume that in the absence of a magnetic field, the specific heat near the critical point diverges as $|\tau|^{-\alpha}$. Then it follows from Eq. (41) that $\psi(+\infty)$ and $\psi(-\infty)$ must remain finite. Since the specific heat is proportional to the second derivative of the free energy with respect to temperature, and so to $\tau^{d/x-2}$, we find that

$$d/x = 2 - \alpha. \tag{2.42}$$

At temperatures below the critical one, a spontaneous magnetic moment M exists. Now M is determined by the derivative of the free energy with respect to the magnetic field h as $h \to 0$. Thus, M will only be nonzero and finite if, as $h \to 0$, $(\partial f/\partial h)_\tau$ becomes independent of h. However,

$$(\partial f/\partial h)_\tau = (-x/y)\tau^{(d/x+1)}h^{-(x/y+1)}\psi'(z), \qquad z = \tau/h^{x/y}. \tag{2.43}$$

Hence, we require that $d\psi/dz$ be proportional to $z^{-(1+y/x)}$ as $z \to -\infty$. As a result, M will be proportional to $\tau^{(d-y)/x}$. Thus, if the spontaneous magnetic moment M tends to zero as $|\tau|^\beta$, when the temperature approaches the critical one from below, we require that

$$\beta = (d - y)/x. \tag{2.44}$$

The magnetic susceptibility χ is, by its definition, proportional to $(\partial M/\partial h)_\tau$ in the limit as $h \to 0$. An analysis similar to that used to derive Eq. (44) shows that this is finite at temperatures below the critical one only if $\psi''(z) \sim z^{-2(1+y/x)}$ as $z \to -\infty$, and then $\chi \sim \tau^{(d-2y)/x}$. Hence, if the susceptibility diverges as $|\tau|^{-\gamma}$, we find that

$$\gamma = (2y - d)/x. \tag{2.45}$$

On the critical isotherm ($\tau = 0$), let the magnetic moment depend on the magnetic field to the power $1/\delta$, so that $M \sim h^{1/\delta}$. In view of (43), $(\partial f/\partial h)_\tau$ is only independent of τ if $\psi'(z)$ is proportional to $z^{-(1+d/x)}$, so that this is how $\psi'(z)$ must behave as $z \to 0$. In that case, $M \sim h^{(d-y)/y}$, so that

$$1/\delta = (d - y)/y. \tag{2.46}$$

The same considerations that have been applied to the free energy density f of Eq. (40) can also be applied to the correlation function

$$G(r, \tau) = \langle [s(0) - \langle s \rangle][s(r) - \langle s \rangle] \rangle. \tag{2.47}$$

The only difference is that G involves a dependence on the distance r, which must be replaced by r/L as we pass from the site lattice to the block lattice.

We consider first the field-dependent sums $h \Sigma s_i$ and $\tilde{h} \Sigma \mu_\alpha$ in the Hamiltonians (36) and (38) of the site and block lattices, respectively. On collecting in these two sums the terms relating to different blocks, and using Eq. (39), we see that

$$hL^d \langle s_i \rangle = hL^y \langle \mu_\alpha \rangle. \tag{2.48}$$

Thus, spins are multiplied by a factor of L^{y-d} on passing from the block lattice to the site lattice. The correlation function $G(r, \tau)$ involves a product of two spins, and so is multiplied by $L^{2(y-d)}$. Hence

$$G(r, \tau) = L^{2(y-d)}G(r/L, \tau L^x). \tag{2.49}$$

In analogy to the derivation of Eq. (41) from (40), we find that Eq. (49) requires

$$G(r, \tau) = r^{2(y-d)}\phi(\tau/r^{-x}). \tag{2.50}$$

We introduce the critical index η which determines the divergence of the correlation function on the critical isotherm ($\tau = 0$) according to $G(r, \tau) \sim r^{-(d-2+\eta)}$ as $\tau \to 0$. Since this limit is independent of τ, it follows from (49) that $\phi(0)$ must be constant, so that

$$d - 2 + \eta = 2(d - y). \tag{2.51}$$

Finally, it follows from Eq. (50) that the correlation length ξ satisfies the equation

$$(\xi/r)^{-x} = \tau/r^{-x}. \tag{2.52}$$

Hence, if we write $\xi \sim \tau^{-\nu}$, we find that

$$\nu = 1/x. \tag{2.53}$$

In Eqs. (42)–(53), we have introduced the critical indices α, β, γ, δ, η, ν, which describe, respectively, the asymptotic behavior near the critical point of the specific heat, spontaneous magnetic moment, magnetic susceptibility, the field dependence of the spontaneous moment along the critical isotherm, the correlation function at the critical temperature, and the correlation length. Of these indices, one (β) is determined only below the critical temperature, and two (δ, η) exist only at the critical temperature. The remaining three indices can be defined for temperatures above the critical one (α, γ, ν) as well as below it (α', γ', ν'). All our analysis applies equally to either side of the critical temperature, so that from scaling theory we predict that

$$\alpha' = \alpha, \qquad \beta' = \beta, \qquad \gamma' = \gamma. \tag{2.54}$$

We have now obtained nine relationships between the critical indices that characterize the thermodynamic behavior of our system near the critical point. These involve the unknown parameters x and y introduced in Eq. (39). On eliminating x and y from these relationships, we obtain the following four formulas relating the critical indices:

$$\alpha + 2\beta + \gamma = 2; \qquad d\nu = 2 - \alpha; \qquad (2 - \eta)\nu = \gamma; \qquad \beta(\delta - 1) = \gamma. \tag{2.55}$$

Equations (54) and (55) contain the main results of the scaling theory of critical phenomena.

The use of dimensional analysis for critical phenomena involves one rather peculiar feature. The only characteristic length of our problem is the correlation length ξ, and this tends to infinity as the critical point is approached. Thus, at the critical point there is no characteristic length at all. Another way of expressing this feature is that in the order–disorder phase transition problem all lengths are equally important, and there is no single dominant characteristic length. Hence, we cannot use scaling to some characteristic dimension of the problem, as in our previous examples. Instead of this, we compare two different problems by scaling theory, and derive some important results from their similarity. Another example of a problem without any characteristic dimension is the theory of turbulence in fluids, while other examples can be found in quantum field theory. We shall return in Section 2.5 to the modern version of scaling theory, which involves the use of renormalization group methods.

2.2   The Derivation of Formulas by Dimensional Analysis

The magnitude of a physical quantity is described by the product of a number and a unit. The basic idea of dimensional analysis is that a physical formula, i.e., a relationship between the magnitudes of physical quantities, must be valid in any system of units. Our analysis in the last section was based on the assumption that this requires physical formulas to be dimensionally homogeneous and we now proceed to prove this hypothesis.

The Π Theorem

An alternative formulation of the basic idea of dimensional analysis, and one which is often more convenient to use in practice, is as follows. Let the functional relationship

$$\phi(x_1, x_2, \ldots, x_n) = 0 \tag{2.56}$$

between the magnitudes x_j of n quantities be valid when the size of the fundamental units is changed in an arbitrary way. Then, if the quantities x_j involve m such units, this equation is satisfied by and equivalent to a functional relationship

$$F(q_1, q_2, \ldots, q_{n-m}) = 0 \tag{2.57}$$

between $(n - m)$ dimensionless variables q_k, a result known as the Π theorem,

To prove this result, we denote the m dimensions by $\alpha_1, \alpha_2, \ldots, \alpha_m$ and write for the dimensions of x_j

$$[x_j] = \prod_{k=1}^{m} \alpha_k^{c_{jk}}. \tag{2.58}$$

If the size of the unit associated with dimensions α_k is divided by a_k, for $k = 1, 2, \ldots, m$, the quantity described in the old units by x_j will be described in the new units by x_j', where

$$\ln x_j' = \ln x_j + \sum_k c_{jk} \ln a_k. \tag{2.59}$$

We now require that $\phi(x_1', x_2', \ldots, x_n') = 0$, and so that $\partial\phi/\partial(\ln a_i)_{a_k=1}$ vanish. Provided that the x_j can be treated as independent variables, we find from Eq. (57) and (59) that, for instance,

$$\frac{\partial \phi}{\partial (\ln a_1)_{a_k = 1}} = \sum_j \frac{\partial \phi}{\partial (\ln x_j')} \frac{\partial (\ln x_j')}{\partial (\ln a_1)} = \sum_j c_{j1} x_j \frac{\partial \phi}{\partial x_j} \qquad (2.60)$$

Let us define new variables y_j, each of degree one in dimension α_1, according to

$$\ln y_j = (1/c_{j1}) \ln x_j. \qquad (2.61)$$

Then it follows from Eq. (60) that

$$\sum_j y_j \frac{\partial \phi}{\partial y_j} = 0. \qquad (2.62)$$

We now define another set of variables,

$$Z_j = y_j/y_n, \qquad (2.63)$$

which are of degree zero in dimensions α_1, i.e., do not involve it, and write

$$\phi(y_1, y_2, \dots, y_n) = \psi(Z_1 y_n, Z_2 y_n, \dots, y_n). \qquad (2.64)$$

Then

$$\frac{\partial \psi}{\partial y_n} = \sum_{j=1}^{n} Z_j \frac{\partial \phi}{\partial y_j} = \frac{1}{Z_n} \sum_j y_j \frac{\partial \phi}{\partial y_j} = 0, \qquad (2.65)$$

in view of Eq. (62). Thus ψ is independent of y_n, and so is a function only of the $n - 1$ variables Z_j, all of which are independent of dimension α_1. The same process can now be repeated with $\partial \psi / \partial (\ln a_2)$ to obtain a function of $n - 2$ variables that are independent of α_1 and α_2, and so on. Eventually, we will obtain a functional relationship of the form of Eq. (57) between $n - m$ dimensionless variables q_r, which is equivalent to our original Eq. (56).

If the variable of interest occurs in only one of the q_j, say q_1, a more convenient form of Eq. (57), entirely, equivalent to it, is

$$q_1 = f(q_2, q_3, \dots, q_m). \qquad (2.66)$$

In particular, if there is only a single dimensionless parameter q_1, Eq. (66) implies that q_1 is equal to a contant.

We note two points in connection with the above analysis. First, in a problem that involves a large number of variables, a solution of the form (57) will contain fewer parameters q_j, and so dimensional analysis will provide more information, if we increase the number of basic dimensions that we use. Thus, while the six dimensions M, L, T, θ, I, and Q which we considered in Section 2.1 are adequate for many problems, it is sometimes useful to introduce extra dimensions. Second, our analysis shows that different basic dimensions correspond to different choices of unit and not necessarily to different types of physical quantity. It is this feature that enables us to increase sometimes the number of basic dimensions that we employ, as shown in the last two problems treated in this section.

Planetary Motion

As our first example of the application of dimensional analysis, we consider the problem of a planet of mass m_1 moving round the sun, whose mass we denote by m_2. Apart from these masses, the parameters involved in a calculation of the orbital period T_0 are the mean distance R of the planet from the sun, and the universal gravitational constant G. This latter is defined by expressing the force between the sun and the planet, when the distance between them is r, as

$$F = G m_1 m_2 / r^2. \tag{2.67a}$$

Hence, in view of Eq. (2),

$$[G] = [F] L^2 M^{-2} = M^{-1} L^3 T^{-2}. \tag{2.67b}$$

From the five parameters T_0, R, m_1, m_2, and G, which involve the three dimensions M, L, T, we can form two dimensionless constants, which can be chosen to be

$$q_1 = R^3 / m_2 T_0^2 G; \qquad q_2 = m_1 / m_2. \tag{2.68}$$

We choose to use m_2, rather than m_1, in q_1 since $G m_2$ determines the force per unit mass on, and so the acceleration of, the planet. Thus, the solution of our problem is of the form

$$R^3 / m_2 T_0^2 G = f(m_1 / m_2) \tag{2.69a}$$

or equivalently

$$T_0 = (R^3 / m_2 G)^{1/2} f_1(m_1 / m_2), \tag{2.69b}$$

where f and f_1 are arbitrary functions. In fact, the full solution is

$$T_0 = 2\pi (R^3 / G(m_1 + (m_2)))^{1/2}, \tag{2.69c}$$

which is of the form (69b) with $f_1 = 2\pi [m_2 / (m_1 + m_2)]^{1/2}$.

Electrical Units

In order to apply dimensional analysis to problems involving electricity, we start by deriving the dimensions of some of the principal quantities involved in these problems. The first such quantity that we consider is the potential difference or voltage V. Since

$$\text{voltage} \times \text{current} = \text{rate of working} = \text{force} \times \text{distance/time}, \quad (2.70a)$$

we find that

$$[V] = I^{-1}ML^2T^{-3}. \tag{2.70b}$$

The dimensions of resistance R, inductance L, and capacitance C can be readily derived from those of voltage, since by their definition

$$[IR] = [L\, dI/dt] = [Q/C] = [It/C] = [V]. \tag{2.71}$$

To derive the dimension of the dielectric constant ε, we make use of Eq. (4),

$$F = q_1 q_2/\varepsilon r^2. \tag{2.4}$$

Hence, since $[q] = [IT]$,

$$[\varepsilon] = I^2 T^2 L^{-2} F^{-1} = I^2 M^{-1} L^{-3} T^4. \tag{2.72}$$

As we showed in Section 2.1, the dimension of the magnetic permeability μ is related to that of ε by $[\varepsilon\mu] = L^{-2}T^2$.

Space-Charge-Limited Currents

An excellent example of the application of dimensional analysis is provided by the problem of space-charge-limited currents; one aspect of this problem was considered in Section 1.5. We consider systems in which a potential difference V_0 is applied between electrodes distance L apart in a medium of dielectric constant ε. We assume that the electrodes supply as many charge carriers, each of charge q and mass m, as are required to provide the current density J, and want to calculate J as a function of the parameters of the system, and especially of V_0 and L. The electric field between the electrodes is not constant because of the existence of space charge, i.e., a volume density of charge carriers which repel other carriers. We consider two different systems. In the first case, we assume that the charge carriers do not undergo any collisions on their passage between the electrodes, so that their motion is determined by the force per unit mass on them, qE/m. This is typical of the situation in a thermionic valve, for instance. Second, we shall consider a system in which the charge carriers undergo so many collisions that they acquire a limiting drift velocity $\bar{v}$ in a field E that is proportional to E. In this case, the motion of the carriers depends on their mobility μ, defined by

$$|\bar{v}| = \mu|E|, \tag{2.73}$$

rather than on q/m. This situation arises typically in insulating solids or liquids that do not contain deep traps, i.e., in the trap-free limit considered in Section 1.5.

Each of these problems involves five variables and the four dimensions M, L, T, and I, so that we expect to find just a single dimensionless variable. In the first case, we write our dimensionless variable q_1 as

$$q_1 = L^a V^b J^c \varepsilon^d (q/m)^h \tag{2.74}$$

and in the second case as

$$q_1' = L^{a'} V^{b'} J^{c'} \varepsilon^{d'} (\mu)^{h'}. \tag{2.75}$$

A solution of Eq. (74) is provided by

$$q_1 = L^{-4} V_0^3 J^{-2} \varepsilon^2 (q/m), \tag{2.76a}$$

so that

$$J = A\varepsilon (q/m)^{1/2} V_0^{3/2} / L^2, \tag{2.76b}$$

where A is a constant, generally of order unity. This result is known as Child's law.

In order to solve the second problem, we note that, from Eq. (73),

$$[\mu] = [v]/[E] = LT^{-1}/[VL^{-1}] = IM^{-1}T^2. \tag{2.77}$$

We then readily find that a solution of (75) is

$$q_1' = L^{-3} V_0^2 J^{-1} \varepsilon \mu, \tag{2.78a}$$

so that

$$J = A' \varepsilon \mu V_0^2 / L^3. \tag{2.78b}$$

In this case, the exact solution, Eq. (1.25), has $A' = \frac{9}{8}$. Thus, while the two problems appear superficially to be similar, dimensional analysis shows that the current density in the two cases depends differently on the potential difference V_0 and on the interelectrode distance L. This result is obtained without our having to set up and solve the full sets of equations (Poisson's equation plus Newton's second law) in each case and shows us the power of dimensional analysis.

One further point is worth noting in connection with this example. In neither system did we introduce all the possible parameters, e.g., q and m separately in the first case or at all in the second case, the area of the electrodes, and their work functions. Instead, we made use of physical knowledge or intuition to choose the relevant parameters that would directly affect the dependence of J on L and V_0, and only included these in our analysis. Such a procedure is vital for making the best use of dimensional analysis.

Vector Lengths

We now turn to a problem for which the ordinary dimensions are not adequate. As a simple example, we consider a projectile fired with initial velocity V at angle α to the horizontal in a gravitational field producing a downward acceleration g, and ask how its range R depends on V and α. If we choose axes with the z axis vertical and so that the motion of the projectile is in the plane $y = 0$, the parameters relevant to our problem can be chosen as R, g, $V_x = V \cos \alpha$, and $V_z = V \sin \alpha$ (see Fig. 2). In terms of M, L, and T, the dimensions of these are L, LT^{-2}, LT^{-1}, and LT^{-1}, respectively. Then ordinary dimensional analysis leads to the solution

$$R = (V_x^2/g)f(V_x/V_z) = (V^2/g)\phi(\alpha), \tag{2.79}$$

say. This tells us that R is proportional to the square of the initial velocity but does not tell us anything about how it depends on the angle of projection α. Such a result is to be expected, since our problem involves four parameters and only two dimensions; hence, according to Eq. (57), its solution involves two dimensionless parameters, in our case V^2/Rg and α or V_x/V_z.

In order to derive the dependence of R on α by means of dimensional analysis, we must increase the number of dimensions that we employ. To do this, we note that distances in directions x and z are quite independent, and there is no essential reason why they should be measured in the same units. Since, as we saw at the beginning of this section, the use of different units is equivalent to using different dimensions, the scalar dimension of length L can be replaced by the vector length dimensions L_x, L_y, and L_z. Such a procedure will be valuable, of course, only in problems that depend anisotropically on distances in different directions.

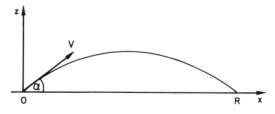

Fig. 2.   The trajectory of a projectile.

For our problem, in terms of the vector lengths, the dimensions of R, g, V_x, and V_z are, respectively, L_x, $L_z T^{-2}$, $L_x T^{-1}$ and $L_z T^{-1}$. Since our problem now involves four parameters and three dimensions, the solution contains a single dimensionless parameter, which we can choose as $Rg/V_x V_z$. Thus, we now find

$$R = AV_x V_z/g = AV^2 \sin \alpha \cos \alpha/g, \qquad (2.80)$$

so that the range is proportional to $\sin 2\alpha$. In fact, for the exact solution, one finds that $A = 2$.

The Thermal Conductivity of a Gas

As a final example of the application of dimensional analysis, we consider how the thermal conductivity K of a gas depends on its molecular mass m, the density of molecules per unit volume N, their mean velocity $\bar{c}$, their mean free path λ, the pressure p of the gas, and its specific heat at constant volume C. This system involves seven parameters, but conventionally only the four dimensions M, L, T, and θ. In order to obtain a unique solution, we require two additional dimensions. We obtain one of these by distinguishing between lengths in the direction, say the x axis, of the thermal gradient, which we denote by L_x, and those perpendicular to it, which we denote by L_0. To introduce another dimension, we note that a distinction can be made between the gravitational mass, for which we denote the dimension by M_g, and the inertial mass M_i. Such a distinction corresponds to associating an independent dimension with the universal gravitational constant G.

We must now derive the dimensions of our parameters. Since only the mean free path and mean velocity in the direction of the temperature gradient, i.e., in the x-direction, are relevant to our problem, $[\lambda] = L_x$ and $[\bar{c}] = L_x T^{-1}$. The quantity of heat Q (which incidentally could have been introduced as an extra dimension, since our problem does not involve the conversion of heat into work) and pressure p are isotropic, and so in their dimensions L is replaced by $(L_0^2 L_x)^{1/3}$, while they also involve the inertial mass M_i. Thus, $[p] = M_i L^{-1} T^{-2} = M_i L_x^{-1/3} L_0^{-2/3} T^{-2}$, while $[Q] = M_i (L_0^2 L_x)^{2/3} T^{-2}$. The specific heat, like the molecular mass, involves gravitational mass, and so $[m] = M_g$, $[C] = [Q/M_g \theta] = M_i M_g^{-1} L_0^{4/3} L_x^{2/3} T^{-2} \theta^{-1}$. The molecular density N is obviously isotropic in distance, so that $[N] = L_0^{-2} L_x^{-1}$. Finally, we note that K is defined by $Q = -KAt \, \partial\theta/\partial x$, where A is area in the plane perpendicular to the x axis, so that $[A] = L_0^2$. Thus, $[K] = M_i L_0^{-2/3} L_x^{5/3} T^{-3} \theta^{-1}$. Elementary but tedious algebra shows that the only dimensionless combination of these parameters is some power of $K^{-1}mN\bar{c}\lambda C$. Then,

$$K = amN\bar{c}\lambda C, \qquad (2.81)$$

where a is some dimensionless constant.

A large number of formulas can be derived by dimensional analysis, especially if one introduces extra dimensions as in the last two examples that we studied. Additional examples will be found in the problem set at the end of this chapter and in the references, while the reader can probably devise others for the problems of specific interest to him.

2.3 Simple Derivation of Physical Laws

In the previous section, we restricted our attention to formulas that arise directly from dimensional analysis, without any assumptions as to the laws of physics that relate the quantities considered. The only need for physical insight in such problems is with regard to the choice of the variables on which the quantity of interest depends. However, if account is taken of the known physical laws, dimensional analysis can be used in a much wider range of problems to derive very simply the functional form of the relationship between different quantities. In this section, we consider two examples of such a procedure. The first of these uses the idea of change of scale which is the basis of dimensional analysis, while the second one involves the use of dimensionless variables.

Motion in a Potential Field

The equations of motion of classical mechanics are linear in the Lagrangian function L, which is the difference between the kinetic energy T and the potential energy U, $L = T - U$ (see Section 3.1). Because of this linearity, multiplication of the Lagrangian by any constant factor does not change the system's equations of motion, and so leaves its properties unaltered. Let us consider a system in which the potential energy U is a homogeneous function of order k of the coordinates $\mathbf{r}_1, \mathbf{r}_2, \ldots, \mathbf{r}_N$ of the particles, so that

$$U(a\mathbf{r}_1, a\mathbf{r}_2, \ldots, a\mathbf{r}_N) = a^k U(\mathbf{r}_1, \mathbf{r}_2, \ldots, \mathbf{r}_N). \qquad (2.82)$$

Specific examples of such systems are the harmonic oscillator, for which $k = 2$, a constant force field (e.g., the gravitational field near the earth) for which $k = 1$, and the Coulomb field or the universal gravitational field, for which $k = -1$. In addition to the transformation of the scale of the coordinates by $r \to ar$, let us change the time scale according to $t \to bt$. As a result of these changes, all velocities will be multiplied by a/b, and the kinetic energy T by $(a/b)^2$. However, the system's properties will remain unaltered only if both terms in L, namely T and U, are multiplied by the same factor. Hence, we require that

$$a^2/b^2 = a^k, \quad \text{i.e.,} \quad b = a^{1-k/2}. \tag{2.83}$$

It follows that for two similar systems, in which the ratios of corresponding lengths and of characteristic times are a and b, respectively, b is related to a by Eq. (83). In other words, the characteristic lengths l, l' and times t, t' of the two trajectories satisfy the equation

$$(t'/t) = (l'/l)^{1-k/2}. \tag{2.84}$$

For the harmonic oscillator, $k = 2$, and so the period of an oscillation is independent of its amplitude. For the free fall of a particle to the earth, $k = 1$ and $(t'/t) = \sqrt{(l'/l)}$, a result which also follows from Eq. (9) since time $=$ distance/velocity. Finally, for the universal law of gravitation, $k = -1$ and $(t'/t) = (l'/l)^{3/2}$. Hence, if the planets rotated about an infinitely massive sun in only its gravitational field, the squares of their periods of revolution would be proportional to the cubes of their average distance from the sun, which is just Kepler's third law. In practice, this law in only approximate, for two reasons. First, the mass of the sun is finite, so that if the sun and planet are regarded as a two-body system the motion of the sun must be taken into account; this leads to the results obtained, from different considerations, in Eqs. (69). Second, the planets move not only in the gravitational field of the sun but also in each other's fields and in those of other celestial bodies, and their distances from each other and from these bodies do not scale with the same factor as their distance from the sun.

Statistical Physics

The use of dimensionless variables provides us with another simple method of deriving the form of physical laws. One field in which this method is especially useful is that of statistical physics, in which the temperature generally appears only in dimensionless variables of the form (energy/$k_B T$), where k_B is Boltzmann's constant. We will first of all present some of the key formulas of statistical physics and then consider examples of their application.

The probability $W_{N,i}$ that a system containing N similar particles is in the quantum state i, with energy $E_{N,i}$, is given by

$$W_{N,i} = \exp[(\Omega + N\mu - E_{N,i})/k_B T], \tag{2.85}$$

where μ is the chemical potential and Ω the free energy. This latter can be found from the condition that the probability is normalized, $\Sigma_{N,i} W_{N,i} = 1$, as a result of which

$$\Omega = -k_B T \ln\left[\sum_N \exp\frac{N\mu}{k_B T} \sum_i \exp\frac{-E_{N,i}}{k_B T}\right]. \qquad (2.86)$$

For an ideal gas of particles, it is convenient to express the energy $E_{N,i}$ of the system and the total number of particles N in terms of the number of particles n_k occupying the single-particle state k with energy ε_k. Since $N = \Sigma_k n_k$ and $E_{N,i} = \Sigma_k n_k \varepsilon_k$, it follows that

$$\Omega = -k_B T \ln \prod_k\left(\sum_{n_k} \exp\frac{n_k(\mu - \varepsilon_k)}{k_B T}\right). \qquad (2.87)$$

Here, the summation over n_k replaces that over N in Eq. (86), while the product over k is just the multiplication of the probabilities of statistically independent events, namely that state k contains n_k particles.

In performing the summation over n_k in Eq. (87), one must take account of the type of statistics that the particles obey. For identical particles obeying Fermi–Dirac statistics, and so subject to Pauli's exclusion principle, $n_k = 0$ or 1. In this case,

$$\Omega_{F-D} = -k_B T \sum_k \ln\left(1 + \exp\frac{\mu - \varepsilon_k}{k_B T}\right). \qquad (2.88)$$

For Bose–Einstein statistics, on the other hand, n_k can be any nonnegative integer, and on summing the resulting geometrical series in Eq. (87) we find that

$$\Omega_{B-E} = k_B T \sum_k \ln\left(1 - \exp\frac{\mu - \varepsilon_k}{k_B T}\right). \qquad (2.89)$$

The average number of particles n_k in a given state k is equal to the derivative of the free energy Ω with respect to ε_k. Thus,

$$n_k = (\exp[(\varepsilon_k - \mu)/k_B T] \pm 1)^{-1}, \qquad (2.90)$$

where the upper and lower signs apply to Fermi–Dirac and Bose–Einstein statistics, respectively. The energy of the one-particle states ε_k is assumed to be known, while for a system containing a fixed number of particles N, the chemical potential μ is determined by

$$\sum_n n_k = \sum_k\left[\exp\left(\frac{\varepsilon_k - \mu}{k_B T}\right) \pm 1\right]^{-1} = N. \qquad (2.91)$$

In general, Ω is a function of T, μ, and either the pressure p or the volume V. However, Eq. (91) shows that if the total number of particles is fixed, the chemical potential is not an independent variable. In such a case, the free energy is identical with the Gibbs free energy F if T and V are the independent variables, and with the Helmholtz free energy, which we denote by Φ, if T and p are chosen as the independent variables.

Equation of State of Fermi and Bose Gases

As an example of the derivation of physical laws by the use of dimensionless variables, we now derive the equation of state for adiabatic processes in Fermi and Bose gases containing a fixed number N of particles. We choose as our independent variables V and T, so that in our case $\Omega = F$.

The first step in our derivation is to pass from a sum over states in Eqs. (88) and (89) to an integral over the energy ε. Let $G(k_0)$ denote the total number of states having wave vector $\mathbf{k}$ with $|\mathbf{k}| \leq k_0$, or equivalently with momentum $p = \hbar k$ no greater than $p_0 = \hbar k_0$. Since k is of dimension L^{-1} while the total number of states $G(k)$ is a dimensionless quantity that is expected to be proportional to the volume V of the system $G(k) = AVk^3$, where A is a dimensionless constant. If the energy of the state k is of the form $\varepsilon(k) = p^2/2m = \hbar^2 k^2/2m$, the total number of states $G(\varepsilon)$ with energy less than ε will then be proportional to $\varepsilon^{3/2}$. Hence, the density of states per unit energy

$$g(\varepsilon) = dG(\varepsilon)/d\varepsilon = aV\varepsilon^{1/2}, \tag{2.92}$$

where a is a constant proportional to $m^{3/2}/h^3$. Moreover, from Eqs. (88) and (89) with $\Omega = F$,

$$F = \mp k_B T \int g(\varepsilon) \ln\left(1 \pm \exp\frac{\mu - \varepsilon}{k_B T}\right) d\varepsilon. \tag{2.93}$$

In order to calculate F, we introduce the dimensionless quantities

$$z = \varepsilon/k_B T, \qquad w = \mu/k_B T. \tag{2.94}$$

We can then write the free energy in the form

$$F = \mp VT^{5/2}k_B^{5/2} \int az^{1/2} \ln[1 \pm \exp(w - z)]\, dz = VT^{5/2}f(w). \tag{2.95}$$

In other words, the free energy per unit volume F/V is a homogeneous function of μ and T of order 5/2, so that a transformation $\mu \to b\mu$ and $T \to bT$

multiplies F by $b^{5/2}$. The derivatives of F with respect to T and μ, which equal the entropy S of the system, and the total number of particles N, are therefore homogeneous functions of μ and T of order $\frac{3}{2}$, so that we can write

$$N = -(\partial F/\partial \mu)_{T,V} = VT^{3/2} f_1(w), \tag{2.96}$$

$$S = -(\partial F/\partial T)_{\mu,V} = VT^{3/2} f_2(w). \tag{2.97}$$

Hence, the entropy per particle S/N is a homogeneous function of zeroth order in μ and T. Thus, for an adiabatic process (S = const) in a system with a fixed number of particles (N = const), $w = \mu/k_B T$ must be constant. In view of Eq. (96), this implies that for adiabatic processes

$$VT^{3/2} = \text{const.} \tag{2.98}$$

Note that we have derived this equation of state without finding the explicit form of the function $f(w)$ that appears in Eq. (95). Moreover, we could not have derived it by the simple methods of Section 2.2, since $VT^{3/2}$ is most certainly not dimensionless.

In some astrophysical problems, a strongly compressible electron gas is considered, in which the average energy of an electron becomes comparable to its rest energy mc^2. In this case, we must use the relativistic formula that connects energy and momentum, $\varepsilon = c\sqrt{m^2c^2 + p^2}$. In the nonrelativistic case, $p \ll mc$, this leads to the usual formula $\varepsilon - mc^2 = p^2/2m$. We now consider the opposite extreme relativistic case, $p \gg mc$, in which case $\varepsilon = pc$. In this case, on transforming from $G(k)$ to $G(\varepsilon)$ we see that $G(\varepsilon)$ is proportional to ε^3 and $g(\varepsilon) = bV\varepsilon^2$. Thus, on substituting in Eq. (93), and using the substitutions of Eq. (94), we find that

$$F = -VT^4 k_B^4 \int bz^2 \ln[1 + \exp(w - z)] \, dz = VT^4 f_r(w). \tag{2.99}$$

Hence, on proceeding as we did from Eq. (95) to (98), we find that the equation of state for adiabatic processes in the ultrarelativistic electron gas is

$$VT^3 = \text{const.} \tag{2.100}$$

Equations (98) and (100) are just two examples of the very common and powerful technique of obtaining the form of physical laws by using dimensionless variables in integrals. Note that the form $F/V = T^n f(\mu/k_B T)$ is reminiscent of the equation of state near the critical point, Eq. (41), used in the scaling theory of phase transitions, and of Eq. (16), which arose in the problem of nonlinear heat conduction.

2.4   Dimensionless Equations and Physical Similarity

The use of dimensional analysis is not restricted to finding the form of the solution of a problem or of the relationship between quantities, as in the examples of Sections 2.2 and 2.3. Another common and very important application is in connection with the quantitative solutions of dimensionless equations. If the equations governing a system's behavior are formulated in terms of the appropriate dimensionless parameters, we often find that a whole range of systems is described by the same equation. In that case, it is only necessary to solve this equation once, numerically if necessary, in order to determine completely the behavior of all the systems. One example of this is provided by the van der Waals equation for a nonideal gas, which as we saw in Section 2.1 can be transformed from a form that depends explicitly on the properties of the specific gas, Eq. (25), to a universal form, Eq. (29), for which the gas' specific properties only determine the relationship between the dimensionless variables and the physical observables. Another example of such a universal equation is the Thomas–Fermi equation for the electron density around an atom, which we consider below. It can also happen that the dimensionless equation contains a parameter which varies from one system to the next, such as Reynolds number [see Eq. (13)] in hydrodynamic problems. This leads to the concept of physically similar or model systems, which we discuss in the second half of this section.

The Electrical Charge Distributions in Atoms—The
Thomas–Fermi Equation

The motion of the electrons in atoms is governed by the attractive Coulomb forces between the heavy nucleus, which to a good approximation can be assumed to have infinite mass, and the light electrons. Thus, the parameters that determine the electron's possible orbits and energies are the electron's mass m, and the charges Ze and $-e$ of the nucleus and the electron, respectively. According to Coulomb's law, these charges will only appear in the combination Ze^2/ε_0, which according to Eq. (4) or (72) is of dimension ML^3T^{-2}. It is well known that classical mechanics cannot explain the stability of atoms. Since classical physics is the limiting case of quantum mechanics when Planck's constant h tends to zero, we expect that h is also a characteristic parameter of our problem; it has the dimension of action, i.e., work times time, $[h] = ML^2T^{-1}$. It can easily be checked that from these parameters, only one characteristic energy E_0 and one characteristic length a can be constructed, and these are given by

$$E_0 = mZ^2e^4/\varepsilon_0^2 h^2, \tag{2.101a}$$
$$a = h^2\varepsilon_0/mZe^2, \tag{2.101b}$$

where we have used $\hbar = h/2\pi$, as a matter of convenience, in place of h. In the exact solution for the ground state of the hydrogen atom, for which $Z = 1$, the electron's orbital has a characteristic radius given by Eq. (101b), and a binding energy differing from that of Eq. (101a) by a factor of $\frac{1}{2}$. Such numerical factors cannot, of course, be derived by dimensional analysis.

For atoms containing more than one electron, an exact solution cannot be found, and some approximate methods are needed. One such method, very suitable for heavy atoms containing many electrons, is the Thomas–Fermi approximations. This is a quasi-classical approximation, in which the different states of the individual electrons are completely neglected. Account is taken of the nonclassical nature of the system by assuming that each cell of phase space of volume h^3 can accommodate only two electrons (one of each spin, in accordance with Pauli's exclusion principle). Thus, the number of electrons per unit volume with momentum p not greater than p_0 is just

$$n = 2(4\pi/3)p_0^3/h^3. \tag{2.102}$$

We denote by $\phi(r)$ the average electrostatic potential at distance r from the nucleus produced by the Coulomb fields of the nucleus and all the other electrons. Thus, the effective potential energy of an electron with orbital angular momentum l is

$$V_{\text{eff}}(r) = -e\phi(r) + \hbar^2(l + \tfrac{1}{2})^2/2mr^2, \qquad l \neq 0. \tag{2.103}$$

The second term in V_{eff} differs from the corresponding term in Eq. (4.6) since in the quasi-classical approximation we must use $(l + \frac{1}{2})^2$, rather than $l(l + 1)$, in the centrifugal energy, if $l \neq 0$. This term in the energy is absent if $l = 0$. Hence, from Eqs. (102) and (103), we see that the maximum total energy of an s-electron (i.e., one for which $l = 0$) at a point r is

$$E_{\text{max}} = p_0^2/2m + V_{\text{eff}}(r) = (3n/8\pi)^{2/3}(\hbar^2/2m) - e\phi(r). \tag{2.104}$$

We note that the quasi-classical approximation that we have used enables us to express the total energy as the sum of a kinetic and a potential energy.

The total energy of an electron bound to an atom must be negative, as otherwise the electron would move off to infinity, and constant at every point in the atom, since otherwise the electrons would move to the point of minimum energy. Thus, the maximum momentum of an electron p_0 is determined by $E_{\text{max}} = 0$, and so if all states with momentum no greater than p_0 are filled, then according to Eq. (104)

$$n(r) = (8\pi/3)(2me\phi)^{3/2}/h^3. \tag{2.105}$$

An additional connection between the electron density $n(r)$ and the potential ϕ is provided by Poisson's equation

$$\nabla^2\phi = en/\varepsilon_0. \tag{2.106}$$

On combining these last two equations, we obtain the Thomas–Fermi equation

$$\nabla^2\phi \equiv (1/r)\, d^2(r\phi)/dr^2 = (8\pi/3)e^{5/2}(2m\phi)^{3/2}/h^3\varepsilon_0. \qquad (2.107)$$

To obtain one boundary condition for this equation, we note that as $r \to 0$, the contribution of the electrons to $\phi(r)$ becomes negligible, so that

$$r\phi(r) \to Ze/\varepsilon_0 \qquad \text{as} \quad r \to 0. \qquad (2.108a)$$

For our other boundary condition, we note that the total number of electrons, and so $\int_0^\infty r^2 n(r)\, dr$, is finite. Hence, in view of Eq. (106),

$$r\phi(r) \to 0 \qquad \text{as} \quad r \to \infty. \qquad (2.108b)$$

We now introduce dimensionless variables in place of r and ϕ. Since r has the dimension of length and $e\phi$ those of energy, the natural choice is to use Eq. (101) and write

$$y = r/a, \qquad f = (r/a)(e\phi/E_0) = r\phi\varepsilon_0/Ze. \qquad (2.109)$$

In terms of these new variables, Eq. (107) and (108) become

$$d^2f/dy^2 = (2\sqrt{2}/3)\pi^{-2}Z^{-1}f^{3/2}y^{-1/2}, \qquad f(0) = 1, \qquad f(\infty) = 0. \qquad (2.110)$$

While Eq. (110) is dimensionless, it still contains the number Z as a parameter, and so has to be solved separately for each type of atom. However, the transformation

$$x = 2(3\pi^2 Z)^{-2/3}y = (2me^2/h^2\varepsilon_0)(3\pi^2)^{-2/3}Z^{1/3}r \qquad (2.111)$$

leads finally to the Thomas–Fermi equation

$$d^2f/dx^2 = f^{3/2}x^{-1/2} \qquad f(0) = 1, \qquad f(\infty) = 0. \qquad (2.112)$$

This equation is independent of the atomic number Z, and so, after it has been solved once numerically, this solution applies to all atoms.

Even without solving the Thomas–Fermi equation explicitly, we can derive some valuable information about the dependence of the solution on Z. We note that, since Z is dimensionless, this information could not be obtained by dimensional analysis alone, without the physical laws expressed by Eqs. (102)–(106). First, the dimensionless parameter x, which is associated with length, is proportional to $Z^{1/3}r$. Hence, if most of the charge is contained in the region with $x < x_0$, the corresponding radius of the atom is proportional to $Z^{-1/3}$, a result that is supported by scattering experiments, as we noted in Section 1.2. Second, the average energy per electron in the atom is proportional to ϕ, and hence, in view of Eqs. (109) and (111), to $Z^{4/3}$. Finally, the total energy of all the electrons in the atom is just Z times this and so is proportional to $Z^{7/3}$. Numerical calculations show that in atomic units, for which the units of length and energy are given by Eqs. (101) with $Z = 1$, $x_0 = 1.33Z^{-1/3}$, and the average electronic energy per atom is $0.76Z^{7/3}$. As expected, both these numerical constants are of order unity.

Another interesting application of the Thomas–Fermi method is to predict the values of Z at which, as the number of electrons in the atom (i.e., Z) increases, orbitals with angular momentum $l = 1, 2,$ and 3 (i.e., p, d, and f orbitals, respectively) start to be occupied. Since ϕ is proportional to $Z^{4/3}$, for small values of Z the second term in the effective potential energy $V_{eff}(r)$ of Eq. (103) may become larger than the first. In that case, V_{eff} will be positive, and hence, since an electron's kinetic energy is nonnegative, E_{max} will also be positive, and the electron will not be bound to the atom. For each value of l, a value Z_{crit} exists such that $V_{eff}(r)$ is always negative if $Z > Z_{crit}$, and the corresponding orbitals will then start being filled. The condition for Z_{crit} is that for it $V_{eff}(r)$ touches the axis $V = 0$, so that $V_{eff}(r) = 0$ and $dV_{eff}/dr = 0$ simultaneously. On solving this equation, once $\phi(r)$ has been found from Eqs. (109), (111), and (112), one finds that

$$Z_{crit} = 0.155(2l + 1)^3. \qquad (2.113)$$

A comparison with the experimental data shows that the numerical factor in Eq. (113) is some 10% too low, but the dependence on l fits the data. Thus, if we replace the factor 0.155 by 0.17, we obtain $Z_{crit} = 4.6, 21.2,$ and 58.3 for $l = 1, 2,$ and 3, respectively, while the observed values are 5 (boron), 21 (scandium), and 58 (cerium). According to this formula, g-electrons ($l = 4$) would be in bound states only for $Z > 124$.

Heat Conduction in a Cubic Block

Another very different problem that can conveniently be treated by the use of dimensionless equations is that of the conduction of heat through a block of material. We consider first a cube of material of side L, initially all at temperature T_0, which is placed at time $t = 0$ inside an enclosure whose temperature is maintained at the constant value T_1. We assume that heat transfer at the block's surfaces is rapid, so that these surfaces can be regarded as being at temperature T_1 for $t > 0$. The rate of conduction of heat from the surfaces to the interior of the block is then the process that determines how rapidly the center of the block warms up (if $T_1 > T_0$) or cools downs (if $T_1 < T_0$). The temperature $T(\mathbf{r}, t)$ then satisfies the three-dimensional heat conduction equation

$$\partial^2 T/\partial x^2 + \partial^2 T/\partial y^2 + \partial^2 T/\partial z^2 = (1/\alpha)\, \partial T/\partial t, \qquad (2.114)$$

subject to the initial and boundary conditions

$$T(x, y, z, 0) = T_0, \qquad 0 \le x, y, z \le L; \tag{2.115a}$$

$$T = T_1 \qquad \text{for} \quad t > 0 \quad \text{on the planes}$$

$$x = 0, \quad x = L, \qquad y = 0, \quad y = L, \qquad z = 0, \quad z = L. \tag{2.115b}$$

The parameter α in Eq. (114) is the material's thermal diffusivity, i.e., its thermal conductivity divided by its specific heat per unit volume.

While a lot of information can be derived just by forming dimensionless variables from the parameters associated with the problem, as in Section 2.2, we are interested now in a complete solution and so prefer to express Eqs. (114) and (115) in terms of dimensionless variables. For lengths, the natural transformation is

$$u = x/L, \qquad v = y/L, \qquad w = z/L. \tag{2.116}$$

In order to give the boundary conditions a form that is independent of the temperatures T_0 and T_1, we replace T by the reduced temperature

$$F = (T - T_1)/(T_0 - T_1). \tag{2.117}$$

With these transformations, Eq. (114) will take the canonical form

$$\partial^2 F/\partial u^2 + \partial^2 F/\partial v^2 + \partial^2 F/\partial w^2 = \partial F/\partial s, \tag{2.118}$$

provided we replace the time t by the variable (which can readily be shown to be dimensionless)

$$s = \alpha t/L^2. \tag{2.119}$$

In terms of our new variables, the boundary conditions become

$$F(u, v, w, 0) = 1, \qquad 0 \le u, v, w \le 1; \tag{2.120a}$$

$$F = 0 \qquad \text{for} \quad s > 0 \quad \text{on the planes}$$

$$u = 0, \quad u = 1, \qquad v = 0, \quad v = 1, \qquad w = 0, \quad w = 1. \tag{2.120b}$$

Thus, the solution of all problems of this type is known once the standard set of Eqs. (118) and (120) is solved.

Equations Involving Parameters

In the above two problems, just as in van der Waals equation for a non-ideal gas, we managed by a suitable change of variables to transform an equation that depended on the system's parameters to one that was independent of these parameters and so is valid for all systems. However, such a transformation is not always possible, and the dimensionless equations that describe a system's behavior often contain one or more dimensionless parameters. For instance, in the heat conduction problem that we have just considered, let us replace our cubic block of material by a rectangular block of sides L, M, and N. If we use the transformations of Eqs. (116), (117), and (119), we shall again obtain Eq. (118), but our boundary conditions will involve the parameters M/L, N/L. The appearance of such parameters in boundary conditions is not very convenient, since it is not easy to assess their relative importance there. Instead, it is preferable to use a different scale factor for lengths in different directions, a procedure analogous to the use of vector lengths at the end of Section 2.2, and to replace Eq. (116) by

$$u = x/L, \qquad v = y/M, \qquad w = z/N. \qquad (2.121)$$

In that case, our boundary conditions will have the canonical form of Eq. (120), but our differential equation becomes

$$\partial^2 F/\partial u^2 + (L/M)^2 \, \partial^2 F/\partial v^2 + (L/N)^2 \, \partial^2 F/\partial w^2 = \partial F/\partial s. \qquad (2.122)$$

Thus, a separate solution is required for each set of parameters $(L/M, L/N)$; of course, the cube is only a special case in which both these parameters are unity.

The use of dimensionless equations containing parameters, such as Eq. (122) has two main advantages. First, while the same solution cannot be used for all systems, it can be used for all systems having the same values of these parameters. Such systems are said to be similar, and this similarity provides the basis for using models and making measurements on them rather than on the original system. For instance, if we wish to study the temperature distribution in an element of a microelectronic circuit of dimensions $0.1 \times 0.2 \times 0.3$ cm, it may well be more convenient to make measurements on a system of dimensions $2 \times 4 \times 6$ cm. This use of models is especially widespread in hydrodynamics, as we discuss below, but is certainly not confined to this field. Second, the use of such an equation frequently enables us to see which terms in the equation are important and which can be ignored in a first approximation. For instance, if in our heat conduction problem L is very much less than M and N, we expect that the terms involving $\partial^2 F/\partial v^2$ and $\partial^2 F/\partial w^2$ will be small and can be ignored, so that we have a one-dimensional diffusion problem instead of a three-dimensional one. This is very reasonable physically since if a thin plate is placed inside a refrigerator, we expect that it will cool down at a rate depending on the conduction of heat perpendicular to its faces, with the temperature fairly uniform in planes parallel to its faces.

Hydrodynamic Modeling

One field in which the governing equations frequently involve dimensionless parameters that cannot be eliminated is hydrodynamics. For instance, in the dimensionless Navier–Stokes equation (32), the Reynolds number, which was defined in Eq. (13), $Re = Ul\rho/\eta$, appears as a dimensionless parameter. While Re could be eliminated from this equation by means of the transformation $\mathbf{r} \rightarrow \mathbf{r}\sqrt{Re}$, i.e., by multiplying $\mathbf{c}$ and $\mathbf{h}$ of Eq. (31) by $\sqrt{Re}$, such a transformation will not be of much use if, as will usually happen, it introduces Re into the boundary conditions.† In that case, Re is an essential parameter of the problem, and this must be solved separately for each value of Re. However, while the same solution cannot be used to all systems, it can be used for all systems having the same Reynolds number, and the behavior of one system deduced from that of another one having the same value of Re.

An essential condition for the use of models is that the model system should behave in exactly the same way as the real one. Let us consider, for instance, experiments to measure the resistance met by a body moving through a fluid. Examples of such systems include the motion of a ship in water and of an airplane in the air. If the fluid could be regarded as incompressible, and the body characterized by a single length l, the behavior of two systems would be similar if they have the same Reynolds number. In that case, we see from Eq. (34), $F = \eta Ul f(Re)$, that the force, measured in units of ηUl, will be the same for both systems. Hence, in order to study a system having a high typical velocity U, a model can be used having a larger l and/or a smaller η/ρ, together with a smaller velocity, so that Re is unaltered, and the forces acting in the original system can then be deduced from those found in the model system. However, in practice the situation is usually more complicated and involves a larger number of characteristic quantities and dimensionless parameters. These parameters are not just arbitrary but determine the relative importance of different effects, just as in Eq. (122), $(L/M)^2$ and $(L/N)^2$ indicated the importance of heat conduction in the y- and z-directions relative to that in the x-direction.

The Reynolds number $Re = Ul\rho/\eta$ could have been derived as the dimensionless ratio of the two forces that appear in the Navier–Stokes equation (1.26). The intertial force $\rho(\mathbf{v} \cdot \nabla)\mathbf{v}$ is typically of magnitude $\rho U^2/l$, and the viscous force $\eta\nabla^2\mathbf{v}$ of magnitude $\eta U/l^2$, and the Reynolds number is just the ratio of these two quantities. Let us now consider a system in which

---

† For an example of a system in which this does not happen, see Problem 9.

the fluid is also exposed to gravity, for instance because it has a free surface. In that case, the gravitational force, of magnitude ρg per unit volume, will enter the Navier–Stokes equation. The ratio of this force to the inertial forces, i.e., $(\rho g)/(\rho U^2/l)$ then determines another dimensionless parameter of the problem, called the Froude number, $\text{Fr} = U^2/lg$. In addition, if the free surface is important, account should be taken of the forces $F = \gamma l$, where γ is the fluid's surface tension. The ratio of the inertial forces to the surface tension forces is the Weber number W; since Eq. (1.26) involves densities of force, i.e., forces per unit volume, $W = (\rho U^2/l)/(\gamma l/l^3) = \rho U^2 l/\gamma$. Hence, in systems for which a free surface is important, care must be taken to ensure that the original and model systems have the same Reynolds, Froude, and Weber numbers.

So far, we have only discussed incompressible fluids. It is well known that in a compressible fluid sound waves can propagate. This leads to another characteristic parameter, the velocity of sound c, which is just the speed of propagation of small adiabatic disturbances, $c = (\partial p/\partial \rho)_S^{1/2}$. Thus a new dimensionless parameter appears, the ratio of the typical velocity U to the speed of sound c, which is known as the Mach number, $M = U/c$.

Fortunately, not all these dimensionless parameters are important in all problems. For instance, compressibility is usually important for gases rather than for liquids. In gases whose density is not too high, the kinetic theory of molecules can be used. According to this, the ratio η/ρ is the product of the molecular mean free path λ and some average velocity v, $\eta/\rho \sim \lambda v$. If we identify v with c and consider a system with the Mach number close to unity, the Reynolds number $\text{Re} \sim Ul/\lambda c = Ml/\lambda \gg 1$. This large Reynolds number means that the viscous forces are small compared with the inertial ones. Hence, the motion of a body through a gas can usually be regarded as motion through an ideal, nonviscous fluid. This is why attention is paid to the Mach number rather than to the Reynolds number in the application of gas dynamics to the design of supersonic airplanes, for instance. However, in such fast motion heat exchange and heat transfer start to play an important role, and other dimensionless parameters that characterize these thermal properties, such as the Prandtl, Grashoff, and Rayleigh numbers, must be introduced.

One can find in the literature more than 200 dimensionless parameters used for modeling in different branches of physics and technology. Each of these is in essence the ratio of two similar quantities (force, pressure, velocity, etc.) and determines which is of greater importance in a specific system. The construction of useful models is only possible because in many cases just one or two of these parameters determine the essential features of a system's behavior.

2.5   Modern Theory of Critical Phenomena

The last type of application of dimensional analysis that we consider is
its use in problems for which no characteristic parameter exists. As we saw
at the end of Section 2.1, one such problem is the behavior of systems close
to the order–disorder phase transition. In this section, we shall again consider an Ising model system near its critical point and describe it in terms
of blocks of spins interacting through an effective Ising-type coupling.
Initially, we will assume that the resulting block lattice is described by a
Hamiltonian of the same form as that for the site lattice, just as we did then.
We then describe the renormalization group method, which enables us to
derive not only scaling relations such as Eqs. (54) and (55) but also explicit
values of the critical indices. On applying this method to a specific system, the
two-dimensional triangular lattice, we will also calculate approximately the
Hamiltonian of the block lattice, by means of the so-called cumulant expansion method. This Hamiltonian is found to be of the same form as that
for the site lattice, so that the hypothesis on which we based all our analysis
is justified. The methods that we describe are now being applied to a very
wide range of problems.

The Ising-type system that we consider has a Hamiltonian H that involves various interactions, and not just those between adjacent spins as in
Section 2.1. We write

$$\beta H = K_1 \sum s_i s_j + K_2 \sum' s_i s_j + K_3 \sum s_i s_j s_k + \cdots, \qquad (2.123)$$

where $\beta = 1/(k_B T)$, T being the temperature, K_1 is the strength of the interaction between spins on adjacent sites, K_2 of that between next nearest-
neighbor sites, K_3 of that between three spins on adjacent sites, etc. The
partition function Z involves a sum over all possible spin configurations
$S = \{s_j\}$,

$$Z = \sum_S \exp[-\beta H(S)]. \qquad (2.124)$$

Hence, both Z and the free energy $F = -k_B T \ln Z$ are determined by the
set of coupling constants $K_\alpha = \{K_1, K_2, K_3, \ldots\}$. For convenience, we have
omitted in our Hamiltonian the possibility of an external magnetic field; such
a field can be regarded as introducing an extra coupling constant K_0 associated with a term Σs_i.

As we discussed in Section 2.1, when the system approaches a critical
point, the correlation length increases without limit, and fluctuations over
all distances contribute to the critical behavior. Hence, we might expect that
critical phenomena will be invariant under the scaling operation

$$r \to r' = r/L. \qquad (2.125)$$

Such a transformation corresponds to that from the site lattice to the block one. For, if r denotes the distance between a spin in block a and a spin in block b in terms of the lattice constant of the site lattice (which we choose as our unit of length), then r' is the distance between the block spins μ_a and μ_b in units of the lattice constant L of the block lattice.

The Renormalization Group

If the critical phenomena are invariant under the scaling operation (125), then when, in the partition function Z, the sum is performed over all the possible internal spin configurations in each block, the effective Hamiltonian that is obtained must be of the same form as the original one. We will show later that this hypothesis is correct. The only difference between the Hamiltonians will be in the coupling constants. The original constants $K_\alpha = \{K_1, K_2, K_3, \ldots\}$ will be replaced by a new set, $K_\alpha^{(1)} = \{K_1^{(1)}, K_2^{(1)}, K_3^{(1)}, \ldots\}$, which are of course functions of the original ones. Thus, the scaling operation (125) is equivalent to a charge in the values of the coupling constant, which we denote by

$$K_\alpha \to K_\alpha^{(1)} = \psi(K_\alpha). \tag{2.126}$$

The scaling operation and associated block construction can be repeated to obtain another set of coupling constants, $K_\alpha^{(2)} = \psi(K_\alpha^{(1)})$, and this process can be repeated indefinitely. These sets of transformations of the effective coupling constants are called the renormalization group. We assume that the function ψ is well defined and is a regular function even in the thermodynamic limit of infinite systems.

In general, a symmetry group is a set of operations that leaves something constant. The renormalization group is so called because, according to our hypothesis, the transformations of Eq. (125) leave unaltered the form of the dependence of the partition function on the system's Hamiltonian. The only effect of such a transformation is to change the coupling constants in accordance with Eq. (126). For a finite system in a space of d dimensions, if the initial site lattice contains N sites, the block lattice will contain NL^{-d} blocks. Since these describe the same problem, we require that their partition functions be equal, i.e., that

$$Z_N(K_\alpha) = Z_{NL^{-d}}(K_\alpha^{(1)}). \tag{2.127}$$

The Gibbs free energy per spin $f(K)$ is related to the partition function Z_N for a finite system according to

$$f(K_\alpha) = -k_B T \lim_{N \to \infty} \frac{1}{N} \ln Z_N(K_\alpha). \tag{2.128}$$

Hence, on applying this formula to the site and block lattices and using Eq. (127), we find that

$$f(K_\alpha) = L^{-d} f(K_\alpha^{(1)}). \tag{2.129}$$

The main idea of the renormalization group theory is that the transformations in this group possess a "fixed point," i.e., that at some stage we reach a set of coupling constants $K_{\alpha*}$ which are unaltered by the transformation, so that

$$\psi(K_{\alpha*}) = K_{\alpha*}. \tag{2.130}$$

In other words, at some stage of the iterative procedure, the transformation of lengths (or scaling) does not change the problem at all. It is natural to identify such a fixed point with the cirictical point.

In order to obtain the critical indices, we consider small deviations from the fixed point. If these are sufficiently small, we can use a linear approximation and write for the individual K_p, using Eqs. (126) and (130),

$$K_p^{(1)} - K_{p*} = \psi(K_p) - \psi(K_{p*}) = A_{pq}(K_q - K_{q*}), \tag{2.131}$$

where a summation over q is implied. An appropriate choice of linear combinations of $K_p - K_{p*}$ enables us to diagonalize the matrix A_{pq}. If we use a prime to denote these combinations and denote the diagonal elements by L^{Z_p}, we can write

$$K_p'^{(1)} - K_{p*}' = L^{Z_p}(K_p' - K_{p*}'). \tag{2.132}$$

In this equation, the indices Z_p determine the critical behavior, and their sign is of crucial importance. If, for a given p, $Z_p < 0$, then after n transformations of type (125) the difference between $K_p'^{(n)}$ and K_{p*}' will be $L^{-n|Z_p|}$ times that between K_p' and K_{p*}'. Thus, this difference decreases exponentially as n increases. Since the critical behavior is determined by long-range effects, which are unaltered by a large number n of transformations (125), we expect that the parameters K_p' with $Z_p < 0$ are irrelevant to the critical behavior. The only important variables will be those for which $Z_p > 0$, so that $|K_p'^{(n)} - K_{p*}'|$ increases as n increases. Such variables are called the relevant variables of the system. (The borderline case of $Z_p = 0$ is more subtle and will not be discussed here.)

For a simple isotropic magnetic system, we expect that there are two relevant variables, namely, the magnetic field h and the reduced temperature $\tau = (T - T_c)/T_c$ which describes the deviation from the critical temperature. Thus, for this problem,

$$K_\alpha = (\tau, h), \qquad K_\alpha^{(1)} = (\tau^{(1)}, h^{(1)}), \tag{2.133}$$

while the critical values are $\tau = 0$ and $h = 0$. Hence, if we denote the associated indices Z_p by x and y, then according to Eq. (132)

$$\tau^{(1)} = L^x \tau, \qquad h^{(1)} = L^y h. \tag{2.134}$$

We then find from Eq. (129) that

$$f(\tau, h) = L^{-d} f(\tau L^x, h L^y). \tag{2.135}$$

This result is just Eq. (40) of Section 2.1, from which we derived both the connections between the critical indices of the system and x and y of Eqs. (42)–(53) and the relations between the critical indices contained in Eqs. (54) and (55). Thus, the renormalization group theory provides a basis for these relationships. In addition, though, it provides us with a means of finding x and y, and hence the values of the critical indices, for a given system if we can find the fixed points of the renormalization group transformations.

An Application of the Renormalization Group Theory

As an example of the use of the renormalization group theory, we now apply it to a very simple problem, that of the two-dimensional triangular Ising lattice, shown in Fig. 3, with only nearest neighbor interactions. In the absence of an external magnetic field, the Hamiltonian H of this system is given by

$$\beta H = -K \sum_{i,j} s_i s_j, \tag{2.136}$$

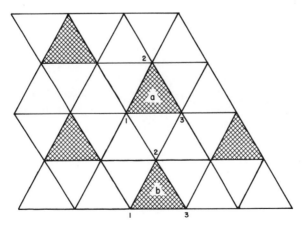

Fig. 3. A triangular lattice for the two-dimensional Ising model; each shaded block contains three spins.

where the sum is over nearest neighbors only. Onsager has obtained the exact solution of this problem, namely, $K_* = \frac{1}{4} \ln 3 \simeq 0.2747$, $x = 1$, and $y = \frac{17}{8}$. The existence of this exact solution is just the reason why it is so instructive to apply to it the renormalization group procedure, since we can then compare our results with the exact ones.

The simplest way to construct blocks in a triangular lattice is to combine groups of three adjacent sites of the original lattice, as shown in Fig. 3. The resulting block lattice, shown by shaded blocks in this figure, is also a triangular lattice, with a lattice parameter corresponding to $L = 3$. We define the spin variable μ_a for the block lattice according to the following rule:

$$\mu_a = \text{sgn}\left(\sum_{i=1}^{3} s_i^{(a)}\right). \tag{2.137}$$

Thus, the block spin μ_a, like the site spin s_i, can take only the values ± 1, and which of these values it takes is determined by the sign of the majority of the site spins in the block. Each value of μ_a corresponds to four different configurations $\sigma_I^{(a)}$ ($I = 1, 2, 3, 4$ or $5, 6, 7, 8$) of the internal spins in the block. For instance, $\mu_a = 1$ is produced by the four configurations

$$\sigma_1^{(a)} = (\downarrow\uparrow\uparrow), \qquad \sigma_2^{(a)} = (\uparrow\downarrow\uparrow), \qquad \sigma_3^{(a)} = (\uparrow\uparrow\downarrow), \qquad \sigma_4^{(a)} = (\uparrow\uparrow\uparrow). \tag{2.138}$$

Since a block can be described either by the $2^3 = 8$ configurations of the site spins $\{s_i^{(a)}, i = 1, 2, 3\}$ or by the variables $\{\mu_a, \sigma_I^{(a)}\}$ with four different configurations as in Eq. (138) for each μ_a, the number of degrees of freedom, in this case eight, is conserved.

In order to obtain the partition function for the block lattice, we just have to sum over all the relevant internal configurations:

$$\exp(\beta H[\mu]) = \sum_{\sigma_I} \exp(\beta H[\mu, \sigma_I]). \tag{2.139}$$

The summation in Eq. (139) can only be performed approximately. In order to carry it out, we rewrite the Hamiltonian H of Eq. (136) as the sum of two terms H_0 and V, which relate, respectively, to the interactions between spins within the same block and those in different blocks. We define the average of a function $f(\mu, \sigma_I)$ with respect to βH_0 by

$$f(\mu) = \langle f(\mu, \sigma_I) \rangle_0 = \sum_{\sigma_I} \exp(\beta H_0[\mu, \sigma_I]) f(\mu, \sigma_I) \bigg/ \sum_{\sigma_I} \exp(\beta H_0[\mu, \sigma_I]). \tag{2.140}$$

Equation (139) can then be written in the form

$$\exp(\beta H[\mu]) = \sum_{\sigma_I} \exp(\beta H_0[\mu\sigma_I]) \langle \exp(\beta V) \rangle_0. \tag{2.141}$$

This formula is still exact. However, we only evaluate $\langle \exp(\beta V) \rangle_0$ approximately, by the so-called cumulant expansion. In this, we write

$$\langle \exp(\beta V) \rangle_0 = \exp\{\langle \beta V \rangle_0 - \tfrac{1}{2}[\langle \beta^2 V^2 \rangle_0 - \langle \beta V \rangle_0^2]$$
$$+ \tfrac{1}{6}[\langle \beta^3 V^3 \rangle_0 - 3\langle \beta^2 V^2 \rangle_0 \langle \beta V \rangle_0 + 2\langle \beta V \rangle_0^3] + \cdots\}.$$

$$(2.142)$$

Since H_0 is the same for every block, if there are N blocks and $Z_0(K)$ is the partition function for one of them, then

$$\sum_{\sigma_I} \exp(\beta H_0[\mu, \sigma_I]) = [Z_0(K)]^N. \qquad (2.143)$$

On substituting from (143) and (142) in (141), we obtain as our final expression for the Hamiltonian of the block lattice

$$\beta H(\mu) = N \ln Z_0(K) + \langle \beta V \rangle_0 - \tfrac{1}{2}[\langle \beta^2 V^2 \rangle_0 - \langle \beta V \rangle_0^2] + \cdots. \qquad (2.144)$$

In order to calculate the averages defined by Eq. (140), we must find the explicit form of the partition function for a single block $Z_0(K)$. To do this, we sum directly over the σ_I of Eq. (138) and obtain

$$Z_0(K) = \exp(3K) + 3\exp(-K). \qquad (2.145)$$

The term $\langle \beta V \rangle_0$ in Eq. (144) connects adjacent spins on different blocks. For instance, for blocks a and b in Fig. 3

$$V_{ab} = K s_2^{(b)}(s_1^{(a)} + s_3^{(a)}). \qquad (2.146)$$

Since the spins $s_2^{(b)}$ and $s_1^{(a)}$ are on different blocks, while the Hamiltonian H_0 neglects the interactions between blocks, on averaging with respect to H_0 we find that

$$\langle s_2^{(b)} s_1^{(a)} \rangle_0 = \langle s_2^{(b)} s_3^{(a)} \rangle_0 = \langle s_2^{(b)} \rangle_0 \langle s_1^{(a)} \rangle_0. \qquad (2.147)$$

We calculate each of these functions directly, using Eqs. (138) and (140). Thus, if for instance $\mu_a = 1$, then

$$\langle s_1^{(a)} \rangle_0 = Z_0^{-1} \sum_{\{\sigma_I\}} s_1^{(a)} \exp(K s_1^{(a)} s_2^{(a)} + K s_3^{(a)} s_1^{(a)} + K s_3^{(a)} s_2^{(a)})$$
$$= Z_0^{-1}[\exp(3K) + \exp(-K)]\mu_a. \qquad (2.148)$$

The reason that in Eq. (148) we have $\exp(-K)$ while in (145) we had $3\exp(-K)$ is that the factor $s_1^{(a)}$, which appears in (148) but not in (146), equals -1 for $\sigma_1^{(a)}$ and $+1$ for $\sigma_2^{(a)}$ and $\sigma_3^{(a)}$, so that the sum of the terms corresponding to these three configurations is just $\exp(-K)$. From Eqs. (148)

and (146), we find that

$$\langle V_{ab} \rangle_0 = 2K \left(\frac{\exp(3K) + \exp(-K)}{\exp(3K) + 3\exp(-K)} \right)^2 \sum{}' \mu_a \mu_b, \qquad (2.149)$$

where the summation is now over blocks. On substituting from (149) in (144), and retaining only terms of first order in V, we finally obtain

$$\beta H[\mu] = K' \sum{}' \mu_a \mu_b, \qquad (2.150)$$

where

$$K' = 2K \left(\frac{\exp(3K) + \exp(-K)}{\exp(3K) + 3\exp(-K)} \right)^2. \qquad (2.151)$$

Thus, in this approximation the Hamiltonian of the block and site lattices, as defined by Eqs. (150) and (136), respectively, do indeed have the same form, but with different interaction constants, which are related by Eq. (151).

We obtain the first-order renormalization equation by setting $K = K' = K_*$ in Eq. (151). The resulting equation has two trivial fixed points, namely, $K_* = 0$ and $K_* = \infty$, which correspond to critical temperature $T_c = \infty$ and $T_c = 0$, respectively. In addition, there is one nontrivial critical point, with

$$K_* = \tfrac{1}{4} \ln(1 + 2\sqrt{2}) = 0.3356, \qquad (2.152)$$

which is not far from the exact result, $K_* = 0.2747$, obtained by Onsager.

In order to find the critical indices, we must find the eigenvalues of the linearized form of Eq. (151) in the vicinity of the fixed point, in accordance with Eqs. (131) and (132). The simplest way to find x is to use the equation

$$[dK'/dK]_{K=K_*} = L^x \qquad (2.153)$$

with K_* given by Eq. (152) and $L = 3$. This leads to the value $x = 0.883$, while Onsager's exact value is $x = 1.000$. The second approximations to $\langle \beta V \rangle_0$ in Eq. (144) is rather more cumbersome and lead to the results shown in Table 1.

To confirm that scaling is of crucial importance in renormalization theory, we now construct another set of blocks on the same initial two-dimensional triangular lattice. Each block now contains seven sites, as shown in Fig. 4. The calculations are carried out in an analogous manner to that used above. The spin variable for a block is again chosen according to the sign of the majority of spins in a block, just as in Eq. (137). However, since there are now seven, rather than three, spins in a block, each value of the block spin corresponds to $2^6 = 64$ configurations, and not just to $2^2 = 4$ as in Eq. (138). The calculations of the partition function for a block, Eq.

Table 1

Values of Critical Point and Index for Block Lattices

	K_*		x	
Exact	0.2747		1.000	
Block lattice	Fig. 3	Fig. 4	Fig. 3	Fig. 4
First approximation	0.3356	0.3003	0.883	0.919
Second approximation	0.2514	0.2647	1.042	1.124
Third approximation	—	0.2752	—	1.072

(145), and the average values $\langle \beta V \rangle_0$ of Eq. (149), $\langle \beta^2 V^2 \rangle_0$, etc., are correspondingly much more cumbersome. The results of the calculations of the critical point K_* and critical index x, up to third order in V, are shown in the table and are quite satisfactory.

Analogous calculations have also been performed for the "magnetic" critical index y by adding a term involving the magnetic field to the Hamiltonian of Eq. (136) and carrying out the analogous renormalization procedure, and satisfactory results are obtained. Results similar to those shown

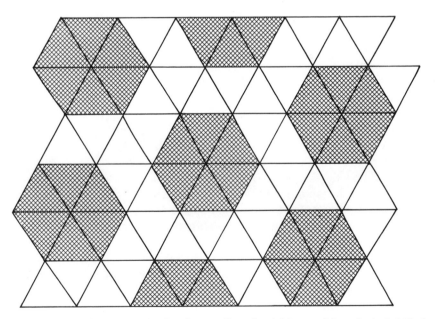

Fig. 4. A triangular lattice for the two-dimensional Ising model; each shaded block contains seven spins.

in the table, and also quite satisfactory, have been obtained for a square lattice by using blocks containing five or nine spins.

Renormalization methods have now been developed and are being applied to a wide range of problems in different fields. We have presented here just one of the simplest examples. This application of dimensional analysis differs from that in the preceding sections of this chapter by applying to a problem in which there are no characteristic parameters whatsoever.

Problems

2.1 The intensity of a monochromatic light wave of cyclic frequency v can be expressed either in terms of its rms electric field E or in terms of its photon density n. Use dimensional analysis to find how E depends on n, Planck's constant h, and the dielectric constant of the medium in which the wave propagates. *Note*: The dimensions of h are such that hv has the dimension of energy. See L. W. Anderson and J. E. Lawler, *Amer. J. Phys.* **46**, 162 (1978).

2.2 The energy density of radiation u in an enclosure in thermal equilibrium at temperature T is found to be proportional to T^4, a result known as the Stefan–Boltzmann Law and which can also be derived from thermodynamics (see Ref. [20, Appendix XXXIII]). Show that no combination of T, Boltzmann's constant k_B, and the velocity of light c can lead to a quantity of the dimensions of u proportional to T^4, so that this result cannot be explained classically, but that if Planck's constant h is introduced, this can be done. Find the form of u for this latter case. See F. Pollock, *Amer. J. Phys.* **40**, 192 (1972).

2.3 Find how the period of a classical particle oscillating in a one-dimensional potential of the form $V = K|x|^s$ depends, for a given arbitrary s, on K, the particle's mass m, and its total energy E. See J. Beyea, *Amer. J. Phys.* **44**, 595 (1976).

2.4 Find how the distance $r(t)$ traveled by a strong shock wave generated by an explosion at time $t = 0$ and the wave's velocity $v(t)$ depend on t, the energy E of the explosion, and the density ρ of the medium in which it propagates, on the assumption that these are the only relevant parameters. See Ref. [13, Section 99].

2.5 A particle of charge q and mass m is released from rest at a point at height h from the ground in a constant horizontal electric field E. By the use of vector dimensions, find how the horizontal distance traveled by the particle depends on q, m, E, h and the gravitational accleration g.

2.6 The density of states function $g(\varepsilon)$ is such that the spatial density of states having energy between ε and $\varepsilon + d\varepsilon$ is $g(\varepsilon)d\varepsilon$. For a quantum-mechanical system, $g(\varepsilon)$ is a function only of ε, the mass m of the particle, and Planck's constant h.

(a) By means of dimensional analysis, find the form of $g(\varepsilon)$ for an n-dimensional system, and compare your results for $n = 3$ with Eq. (92).

(b) For a two-dimensional system of fermions, calculate the Fermi energy $\varepsilon_F(T)$ for a system containing n_0 particles per unit area; note that the integrals can be evaluated exactly.

(c) For a two-dimensional system of bosons, calculate how the Bose energy ε_b, defined analogously to the Fermi energy by expressing the probability $f(\varepsilon)$ that a state of energy ε be occupied in the form $f(\varepsilon) = \{\exp[(\varepsilon - \varepsilon_b)/k_B T] - 1\}^{-1}$ depends on the particle density n_0, and show that this equation can be satisfied for all temperatures. We note that this is not the case for a three-dimensional system, and this leads to the phenomenon of Bose condensation in three dimensions. See J. P. McKelvey and E. F. Pulver, *Amer. J. Phys.* **32**, 749 (1964).

2.7 A finite velocity of propagation for temperature changes can result not only from nonlinearity, as discussed in Section 2.1, but also from a delayed response of the heat current to a temperature change, so that the heat conduction equation for a one-dimensional system becomes

$$C \, \partial\theta/\partial t = \frac{\partial^2}{\partial x^2} \int_{-\infty}^{t} K(t - t')\theta(x, t') \, dt',$$

where C is the specific heat per unit length. A specially useful form of $K(t - t')$ is

$$K(t - t') = (K_0/\tau) \exp[-(t - t')/\tau].$$

(a) By differentiating this heat conduction equation, with the above form for $K(t - t')$, show that in this case

$$C \, \partial^2\theta/\partial t^2 + (C/\tau) \, \partial\theta/\partial t = (K_0/\tau) \, \partial^2\theta/\partial x^2.$$

(b) Choose dimensionless variables $s = t/\tau$ and $y = Ax$, where A is to be determined as a function of C, K_0, and τ, so that our equation becomes the dimensionless telegraph equation for $\theta(y, s)$. This is another example of an equation without parameters that can be used to describe a whole series of physically similar systems. See R. J. Swenson, *Amer. J. Phys.* **47**, 76 (1978).

2.8 In Section 2.4, we solved the Thomas–Fermi equation for a neutral atom, where the boundary condition on the potential was $r\phi(r) \to 0$ as $r \to \infty$, as stated in Eq. (108b). For an ion having charge ze, this boundary condition is no longer correct. Instead, we define an ionic radius R such that outside the sphere of this radius, the electric field is that of a free ion, while on its surface an electron has zero potential energy, so that bound electrons cannot leave this sphere. Thus, we require that, instead of (108b),

$$\phi(R) = 0, \qquad (d\phi/dr)_R = [(d/dr)(ez/r)]_R = -ez/R^2.$$

Define a quantity X that is related to R in the same way as x is to r and show that the boundary conditions for the Thomas–Fermi equation (112) become in this case

$$f(0) = 1, \qquad f(X) = 0, \qquad (x\,df/dx)_X = -z/Z.$$

Note that the problem, and hence the solution, is now characterized by an additional parameter, namely the ratio z/Z of the ionic to the nuclear charge. Thus, this parameter plays a role similar to that of the Reynolds number in hydrodynamic problems, except that here it appears in the boundary conditions. Show that, for $z = 0$, the above boundary conditions imply that X is infinite, so that we regain our original boundary condition, $f(\infty) = 0$. (*Hint*: If X is not infinite, $f(X) = f'(X) = 0$.) In other cases, show that, for given z/Z, the ionic radius R is proportional to $Z^{-1/3}$. Why is this result not of very much practical use? See Ref. [20, Chapter VI, Section 9].

2.9 One of the few problems in hydrodynamics for which the equation of motion for the fluid and the boundary conditions can be put into a dimensionless form that does not involve a parameter such as the Reynolds number Re is the following one, which is known as Karman's problem. An incompressible fluid of density ρ and viscosity η occupies the half-space $z > 0$ and is set in motion by an infinite flat disk in the plane $z = 0$, which rotates about the z-axis with the steady angular velocity Ω.

(a) From the Navier–Stokes equation (1.26) and the continuity equation (1.27), derive equations, in cylindrical polar coordinates (r, ϕ, z), for the components of velocity in these coordinates, v_r, v_ϕ, and v_z.

(b) By means of the change of variables: $s = (\rho\Omega/\eta)^{1/2}z$,

$$p' = -\eta\Omega p, \qquad v_r = r\Omega F(s), \qquad v_\phi = r\Omega G(s), \qquad v_z = (\eta\Omega/\rho)^{1/2}K(s),$$

show that the hydrodynamic equations, together with the boundary conditions for this problem, (namely $v_r = v_z = 0$ and $v_\phi = \Omega r$ on plane $z = 0$; $v_r \to 0$, $v_\phi \to 0$, and $v_z \to$ const as $z \to \infty$) become a set of dimensionless equations and boundary conditions that do not involve any parameters. Thus, if they are solved once numerically, the solution will apply to all systems of this type. See Ref. [13, Section 23].

2.10 Apply the renormalization group method to calculate the partition function Z for the one-dimensional Ising model, for which the total energy for a system of N spins is given by $E = -J\sum_{n=1}^{N} \sigma_n \sigma_{n-1}$, where each spin σ can equal ± 1 and $\sigma_0 = \sigma_N$. *Hint*: In calculating Z, sum first the expression for Z over all the "even" spins σ_{2r}, $1 \leq r \leq N/2$, to obtain an expression for Z involving only the odd spins. This expression will involve terms of the form $\exp[K(\sigma_1 + \sigma_3)]$ $+ \exp[-K(\sigma_1 + \sigma_3)]$, where $K = J/k_B T$, and by writing these in the form $A \exp(K'\sigma_1\sigma_3)$, and finding A and K' as functions of K, you can obtain an expression for Z for the "block" lattice. See H. J. Maris and L. P. Kadanoff, *Amer. J. Phys.* **46**, 652 (1978).

Chapter 3 | Symmetry

3.1 Introduction

Symmetry is one of the most unexpected and fascinating features of the universe. We take pleasure and find peace of mind in the symmetry of the motion of the celestial bodies, in the symmetric geometry of crystals, in symmetries in the vegetable and animal kingdoms, and also enjoy the symmetry of works of art, which is affected by and associated with that of the surrounding world. It is true that art, especially painting, has recently tried to abandon symmetry, apparently in protest against the traditional tenor of life. Fortunately, however, even modern techniques cannot easily change the symmetry of the world around us and of the laws describing the processes that occur in it. Therefore, the study of symmetry, and of its uses in learning about the laws of nature, is very important.

The simplest definition of symmetry is that given by Herman Weyl, and it is almost self-evident. According to his definition, an object (or a physical law) is symmetric if it is possible to do something with it such that, after this operation, the object (or the equation describing a physical law) remains the same as it was originally. As we shall see later, symmetry plays two main roles in physics. First, the symmetry of physical systems with respect to different operations gives us an insight into the origin of such surprising phenomena as the laws of conservation of energy, momentum, electrical charge, etc. Such an explanation is very important both for the completeness of our picture of the world and for our internal satisfaction. Section 3.2 is devoted to a discussion of these matters. Second, the use of symmetry properties both stimulates and greatly simplifies the performance of physical experiments and the description and systematic analysis of their results.

Some physical phenomena, such as the order–disorder phase transitions for instance, are determined and describable almost entirely in terms of symmetry. In Section 3.3, we consider the application of symmetry to the microscopic properties of systems, while Section 3.4 describes its applications to macroscopic properties. We will now describe and discuss these topics in more detail.

Classical Mechanics

The behavior of a mechanical system is determined by the dependence on time of the coordinates describing its arrangement in space. The equations of classical mechanics involve as variables time, the spatial coordinates, and the derivatives of the coordinates with respect to time, i.e., velocity and acceleration. Therefore, all the symmetry properties of mechanical phenomena can be associated with different types of transformation of the spatial coordinates and the time.

Let us start with the purely geometric types of symmetry, which reflect the general properties of the homogeneity of space and time and the isotropy of space. This homogeneity means that there are no fixed reference points in space and that there is no preferred instant of time. In other words, a displacement in space of the system as a whole or a shift in time will not change the mechanical properties of a closed system (i.e., one that does not interact with other systems). In addition, the isotropy of space means that all directions in space are equivalent, so that rotation in space does not change the properties of a closed system.

The mathematical description of these almost self-evident symmetry properties of space and time will be considered here in the framework of Lagrangian mechanics. In this, a closed mechanical system is described by the so-called Lagrangian L, a function that depends on the coordinates $\mathbf{r} = (\mathbf{r}_1, \mathbf{r}_2, \ldots, \mathbf{r}_n)$ and velocities $\dot{\mathbf{r}} = (\dot{\mathbf{r}}_1, \ldots, \dot{\mathbf{r}}_n)$ of all the n discrete particles in the system under consideration, $L = L(\mathbf{r}, \dot{\mathbf{r}})$. The laws of motion of the particles are given by Lagrange's equations

$$\partial L/\partial \mathbf{r}_i - (d/dt)(\partial L/\partial \dot{\mathbf{r}}_i) = 0 \qquad (3.1)$$

and so are uniquely determined by the Lagrangian. Hence the effects of a symmetry operation on the system's equations of motion can be determined from its effect on the Lagrangian.

Symmetry with respect to a displacement of the system in space means that the Lagrangian L is unchanged, i.e., $\delta L = 0$ if every particle in the system is displaced by the same infinitesimal amount $\delta \mathbf{r}$. Thus, in view of (1),

$$\delta L = \sum_{i=1}^{n} \frac{\partial L}{\partial \mathbf{r}_i} \cdot \delta \mathbf{r} = \delta \mathbf{r} \cdot \frac{d}{dt}\left(\sum_{i=1}^{n} \frac{\partial L}{\partial \dot{\mathbf{r}}_i} \right) = 0. \qquad (3.2)$$

Hence, in a closed mechanical system, throughout the motion the vector

$$\mathbf{P} = \sum_i \frac{\partial L}{\partial \dot{\mathbf{r}}_i} = \sum_i \mathbf{p}_i \tag{3.3}$$

remains constant. Thus, the law of conservation of $\mathbf{P}$, which is defined to be the momentum of the system, is a consequence of the spatial homogeneity of the system. Since the Lagrangian is the difference between the kinetic and potential energy of the system,

$$L = \sum_{i=1}^{n} \frac{m_i \dot{\mathbf{r}}_i^2}{2} - U(\mathbf{r}), \tag{3.4}$$

the above definition of the momentum agrees with the usual one.

Homogeneity in time means that the Lagrangian of a closed system does not depend explicitly on the time t. Thus,

$$\begin{aligned}
\frac{dL}{dt} &= \sum_{i=1}^{n} \left(\dot{\mathbf{r}}_i \cdot \frac{\partial L}{\partial \mathbf{r}_i} + \ddot{\mathbf{r}}_i \cdot \frac{\partial L}{\partial \dot{\mathbf{r}}_i} \right) \\
&= \sum_{i=1}^{n} \left(\dot{\mathbf{r}}_i \cdot \frac{d}{dt} \frac{\partial L}{\partial \dot{\mathbf{r}}_i} + \ddot{\mathbf{r}}_i \cdot \frac{\partial L}{\partial \dot{\mathbf{r}}_i} \right) = \frac{d}{dt} \left(\sum_{i=1}^{n} \dot{\mathbf{r}}_i \cdot \frac{\partial L}{\partial \dot{\mathbf{r}}_i} \right).
\end{aligned} \tag{3.5}$$

Hence, another quantity that is constant throughout the motion is the energy E, defined by

$$E = \sum_{i=1}^{n} \dot{\mathbf{r}}_i \cdot \frac{\partial L}{\partial \dot{\mathbf{r}}_i} - L. \tag{3.6}$$

The equivalence of this definition of the energy and its usual definition as the sum of the kinetic and potential energy follows immediately from Eqs. (4) and (6).

We now turn to the isotropy of space, as a result of which the Lagrangian is not changed by a rotation of the system as a whole. We define an infinitesimal rotation by the vector $\delta\boldsymbol{\phi}$, whose length is the angle of rotation and whose direction is the axis of rotation. As shown in Fig. 1, the displacement $\delta\mathbf{r}_i$ of particle i as a result of this rotation is just

$$\delta\mathbf{r}_i = \delta\boldsymbol{\phi} \times \mathbf{r}_i. \tag{3.7a}$$

Similarly, the velocities are changed by

$$\delta\dot{\mathbf{r}}_i = \delta\boldsymbol{\phi} \times \dot{\mathbf{r}}_i. \tag{3.7b}$$

Hence, by calculations analogous to those used for Eqs. (2) and (6), we find that

$$\delta L = \sum_i \left(\delta\mathbf{r}_i \cdot \frac{\partial L}{\partial \mathbf{r}_i} + \delta\dot{\mathbf{r}}_i \cdot \frac{\partial L}{\partial \dot{\mathbf{r}}_i} \right) = \delta\boldsymbol{\phi} \cdot \frac{d}{dt} \sum_i (\mathbf{r}_i \times \mathbf{p}_i) = 0. \tag{3.8}$$

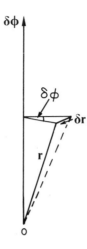

Fig. 1. The displacement $\delta \mathbf{r} = \delta \boldsymbol{\phi} \times \mathbf{r}$ corresponding to a rotation $\delta \boldsymbol{\phi}$.

We thus obtain the conservation law for the vector

$$\mathbf{M} = \sum_i (\mathbf{r}_i \times \mathbf{p}_i), \tag{3.9}$$

which is called the angular momentum of the system.

Thus, all the main conservation laws of mechanical quantities (energy, linear momentum, angular momentum) are an immediate consequence of the general symmetry properties of time and space. Our analysis also explains why the following pairs of variables are associated with each other:

$$\mathbf{r} \leftrightarrow \mathbf{p}, \qquad \boldsymbol{\phi} \leftrightarrow \mathbf{M}, \qquad t \leftrightarrow E. \tag{3.10}$$

Frames of Reference and Relativity

When discussing the homogeneity and isotropy of space, we compared two states of a mechanical system that differed by a displacement or rotation of the entire system. The same properties can, of course, be derived by considering the same mechanical system in two different frames of reference, one of which has its origin displaced or system of axes rotated relative to that of the other. The homogeneity and isotropy of space ensure that mechanical processes are the same in the two frames, and this leads, as previously, to the conservation laws for linear and angular momentum.

However, it is found that the equations of mechanics, and hence the phenomena that they describe, are also unchanged if one frame of reference moves relative to the other with a constant velocity $\mathbf{v}$. In other words, the equations of mechanics are invariant under the so-called Galilean transformation

$$\mathbf{r}' = \mathbf{r} + \mathbf{v}t, \qquad \mathbf{v} = \text{const.} \tag{3.11}$$

This equivalence of all inertial frames† moving uniformly in a straight line relative to each other, which is known as the Galilean relativity principle, can be regarded as expressing a symmetry with respect to uniform rectilinear motion.

Over a hundred years ago, it was found that Maxwell's equations for the electromagnetic fields are not invariant under the Galilean transformation of Eq. (11). As is well known, this does not mean that the principle of relativity is broken in electrodynamics. Rather, this principle holds for the more general Lorentz transformation,

$$x = (x' - vt')/(1 - v^2/c^2)^{1/2}, \qquad y = y',$$
$$z = z', \qquad t = (t' - vx'/c^2)/(1 - v^2/c^2)^{1/2}, \tag{3.12}$$

for frames with parallel sets of axes whose relative direction of motion is parallel to the x axis. The Galilean transformation of Eq. (11) is just a limiting case of this for velocities v much less than the velocity of light c.

Equation (12) forms the foundation of the special theory of relativity. In all our discussion up to now, we assumed that obviously $t' = t$, i.e., the course of time is unaffected by a change in the frame of reference. However, we see from Eq. (12) that for the same instant of time t in one frame, the time t' in another frame depends on $\mathbf{r}'$. Thus, the concept of the simultaneity of two events taking place at different points in space depends on the choice of the frame of reference. In other words, we cannot consider separately the symmetry properties of space and time, as we did previously. As can readily be checked from Eq. (12), the quantity

$$s^2 = x^2 + y^2 + z^2 - c^2t^2 \tag{3.13}$$

is a constant in all inertial frames. The first three terms in this expression for s^2 are the square of the length of the vector with coordinates (x, y, z), while s can be regarded as the length of the 4-vector (x, y, z, ict). The concept of a 4-vector is often used in the special theory of relativity. For instance, a conservation law exists for the 4-vector momentum–energy. Since only space–time is homogeneous, and not space and time separately, momentum and energy are not conserved separately, but only a combination of them similar to s^2. As noted previously, the Lorentz transformation (12) reduces to the

---

† In all such frames of reference, the law of inertia holds, i.e., a moving body on which no forces act will continue indefinitely to move with uniform velocity, rather than come to rest. In fact, this is one of the most surprising laws of mechanics, and there is no common-sense reason why it should be true.

Galilean one of Eq. (11) for velocities considerably less than the velocity of light, and we then return to the symmetry properties and resultant conservation laws considered at the beginning of this section.

Quantum Mechanics

A quantum-mechanical approach to conservation laws is also possible, and again leads to the conservation of energy, momentum, and angular momentum, as we discuss in Section 3.2. In addition, we shall see there that the existence of symmetry with respect to inversion, i.e., a simultaneous change of the signs of all coordinates, leads in the framework of quantum mechanics to a conservation law for parity. This is in contrast to classical mechanics, where such a symmetry does not lead to new conservation laws. Another peculiarity of the quantum-mechanical approach is the existence of the indistinguishability principle for similar particles, and the associated symmetry properties with respect to permutations of the particles' coordinates. According to such a symmetry, all particles can be divided into two groups, having integer or half-integer values of the spin and obeying Bose–Einstein or Fermi–Dirac statistics, respectively.

*Classical Electrodynamics

New conservation laws appear when we go beyond purely mechanical phenomena and consider classical electrodynamics. The interaction of charged particles with an electromagnetic field can be described by adding to the Lagrangian density $\mathscr{L}_0$ of the noninteracting system a term $\mathscr{L}_{int}$, i.e.,

$$\mathscr{L} = \mathscr{L}_0 + \mathscr{L}_{int}; \qquad \mathscr{L}_{int} = \frac{1}{c} \sum_{\mu=1}^{4} j_\mu A_\mu. \dagger \tag{3.14}$$

Here, we have introduced two 4-vectors, namely, the electrical current density j and the vector potential A of the electromagnetic field. The three spacelike components of j describe the current density of charged particles, while the timelike component is associated with the charge density, $j_4 = ic\rho$. The spacelike components of A are just the vector $\mathbf{A}$, which determines the magnetic field according to $\mathbf{B} = \nabla \times \mathbf{A}$, while the timelike component $A_4 = i\phi$ is associated with the electric field $\mathbf{E} = -\nabla\phi - (1/c)\,\partial\mathbf{A}/\partial t$. From

---

† Henceforth, we will use the summation convention, according to which summation is implied over any index that appears twice in one term.

this definition of A, we see that the 4-vector A is only defined to within a so-called gauge transformation of the form

$$A_\mu \to A'_\mu = A_\mu + \partial\theta/\partial x_\mu, \tag{3.15}$$

where θ is an arbitrary function. Such a transformation does not change the observable magnitudes of the electric and magnetic fields $\mathbf{E}$ and $\mathbf{H}$.

The interaction term $\mathscr{L}_{\text{int}}$ in Lagrangian density (14) is altered in the following way by the gauge transformation:

$$\begin{aligned}
\mathscr{L}_{\text{int}} \to \mathscr{L}'_{\text{int}} = j_\mu A'_\mu &= \mathscr{L}_{\text{int}} + (1/c)j_\mu \partial\theta/\partial x_\mu \\
&= \mathscr{L}_{\text{int}} + (1/c)(\partial/\partial x_\mu)(j_\mu \theta) - (1/c)\theta \, \partial j_\mu/\partial x_\mu.
\end{aligned} \tag{3.16}$$

Since the electromagnetic interaction relates to observable magnitudes, it cannot be changed by a gauge transformation. In order to obtain the ordinary Lagrangian L, we must integrate the Lagrangian density $\mathscr{L}$ over the spatial variables x_1, x_2, and x_3. The integral of $(1/c)(\partial/\partial x_\mu)(j_\mu \theta)$ has the form df/dt, the complete differential with respect to time of some function, and it is a well-known property of the Lagrangian that the addition to it of such a term does not affect the system's equations of motion. However, in order to ensure that the last term in Eq. (16) does not affect the system's properties, we must require that $\partial j_\mu/\partial x_\mu = 0$. This is just the charge conservation equation

$$\partial\rho/\partial t + \nabla \cdot \mathbf{j} = 0, \tag{3.17}$$

since $x_4 = ict$.

We see that this equation results from the symmetry of the electromagnetic equations with respect to gauge transformations, just as the conservation laws of mechanics arise from symmetry with respect to Galilean or Lorentz transformations.[†] In Section 3.2, we shall derive this conservation law for electrical charge in the framework of quantum field theory.

Elementary Particles

All the conservation laws that we have considered so far have been derived from the symmetries of a known Lagrangian that describes the properties of the system. However, when we turn to the atomic nucleus and elementary particles, we are confronted with systems whose Lagrangian is not known. In such a case, it is natural to assume that the observed conserva-

---

† The law of conservation of electric charge explains, in particular, the stability of the electron. Since it possesses the least mass of all the charged particles, and charge must be conserved, it has no tendency to break down into any other charged particles. (Compare, however, the discussion of quarks in Section 1.3.)

tion laws, such as the conservation of baryon and lepton charges and of strangeness, are the consequences of some as yet unknown symmetries. In this case, a study of the conservation laws can still enable us to predict properties of these particles, as well as providing restrictions as to the possible form of the system's Lagrangian.

As an example of the conservation laws for baryon and lepton charges, let us consider β-decay, one of the best known nuclear transformations, in which a radioactive nucleus emits an electron and an antineutrino, or a positron and a neutrino, and is thereby transformed into another nucleus whose charge differs from that of the original one by a unit charge. In one such transformation, one of the nuclear neutrons (n) transforms into a proton (p), emitting an electron (e) and an antineutrino ($\tilde{\nu}$), i.e.,

$$n \rightarrow p + e + \tilde{\nu}. \tag{3.18}$$

This reaction is very interesting from both a philosophical and a historical point of view. Philosophically, the appearance as a result of the transformation of an electron that was not initially present in the nucleus makes us revise the commonsense concept that "elementariness" is obtained by dividing the whole into its component parts. Historically, an analysis of this reaction led Pauli, in 1930, to predict the existence of the neutrino, which is very difficult to observe directly. Since the three known particles (n, p, e) taking part in the reaction all have half-integral spin, the total spin of the two sides of Eq. (18) will only be the same if another particle of half-integral spin, which he called the neutrino, is present. This new particle cannot carry any electric charge, since such charge must be conserved in the reaction. Another of its properties can be derived from the conservation of baryon and lepton charge. As we saw in Section 1.3, two of the particles taking part in the reaction, n and p, are baryons, and Eq. (18) contains just one of these on each side. Otherwise, when a baryon appears as a result of a nuclear reaction, it can only be produced along with an antibaryon. The application of similar considerations to the lepton charge leads us to the conclusion that the neutrino, like the electron, must belong to the group of leptons and that one of these particles is a lepton and the other an antilepton.

An analysis of the β-decay reaction, Eq. (18), led Yukawa in 1935 to propose that the nucleons (neutrons and protons) in a nucleus interact with each other by the exchange of mesons, just as the interaction between electrical charges can be ascribed to an exchange of photons. The resulting theory of mesodynamics is not nearly as well developed as electrodynamics, but recent progress in this direction was mentioned at the end of Section 1.3. Any postulated Lagrangian for this interaction must, of course, have symmetry properties that lead to a law of conservation of baryon charge, just as $\mathscr{L}_{int}$ of Eq. (14) led to the conservation of electric charge.

Although its source in the symmetry properties of the system is unknown, the conservation law for baryon and lepton charges at least has the same form as that for electrical charges. A much more peculiar law is the conservation of "strangeness," a quantum number that characterizes particles with anomalously long lifetimes (10^{-7}–10^{-10} sec instead of around 10^{-23} sec). Experiments carried out in the 1950s showed that strangeness is conserved only in processes involving just strong and electromagnetic interactions, and it is not necessarily conserved in processes such as β-decay that involve weak interactions. The breakdown of the strangeness conservation law in this case is accompanied by another phenomenon which was discovered by an analysis of the angular distribution of the electrons in the same β-decay. The essence of Wu's (1957) classical experiment is shown schematically in Fig. 2.

The sample exhibiting β-decay is located in the magnetic field **H** of a circular current. This system is obviously symmetric with respect to reflection in the plane of the current (mirror symmetry), since the magnetic field **H** is an axial vector whose direction is not changed by such a mirror reflection. However, a sharp asymmetry was discovered experimentally: about 40% more electrons were emitted toward one side of the plane than toward the other side. This shows that the particles themselves do not possess mirror symmetry, but are affected by a change of left to right, or magnetic north pole to south pole. Instead of talking about symmetry with respect to mirror reflection, we can talk about symmetry with respect to inversion, $\mathbf{r} \to -\mathbf{r}$, i.e., about parity, since inversion is just a mirror reflection in an arbitrarily oriented plane followed by a rotation of 180° about an axis perpendicular to the plane.

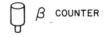

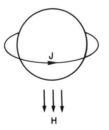

Fig. 2. Wu's classical β-decay experiment (schematic); the circle denotes a sample containing ^{60}Co. [See C. S. Wu *et al.*, *Phys. Rev.* **105**, 1413 (1957).]

The nonconservation of parity in weak interactions such as β-decay is a very surprising result, as it contradicts the seemingly obvious equivalence of left and right. In 1956, Lee and Yang showed that in the weak interactions there is a more general symmetry, in which inversion is combined with the replacement of each particle by its corresponding antiparticle. Neither of these operations changes the properties of strong and electromagnetic interactions, but for weak interactions only their combination is a symmetry operation. In Section 3.2, we shall return to a detailed analysis of the connection between symmetry and the conservation laws.

Molecular Vibrations

The symmetries that we have considered so far have been very general ones, such as the homogeneity of time and space. We now turn our attention to the symmetry properties of specific systems, which have widespread implications for both their microscopic and their macroscopic properties. One typical microscopic property, that we consider first, is the nature and properties of a system's normal modes of vibration.†

The problem that we consider is the effect of rotational symmetry on a system's vibrations. This is most readily demonstrated for small molecules, and we choose as an example the vibrational modes of a carbon dioxide molecule. In equilibrium, this is a linear molecule, with the carbon atom situated at a point A midway between the two oxygen atoms, as shown in Fig. 3. The symmetry operations that leave this molecule unchanged are reflection in a plane through A perpendicular to the molecule's axis, which is conventionally denoted by σ_v, and an infinite set of rotations about this axis.

A general vibration of the molecule, such as that shown in Fig. 3a, has no particular symmetry, as each of the molecule's symmetry operations (apart from the identity) changes it into a different, linearly independent vibration. However, there are certain sets of vibrations which are converted into themselves or into each other by the system's symmetry operations, and we expect the normal modes to be of this type. A proof of this postulate is indicated in Section 3.3. We restrict our attention to internal vibrations of the molecule, i.e., to those motions of the atoms that do not displace the center of mass nor correspond to a rotation of the molecule as a whole. A molecule containing n atoms, each of which can move in three independent directions, possesses $3n$ degrees of freedom. Of these, 3 correspond to the motion of the molecule's center of mass, and for a general molecule another

† The application of a crystal's translational symmetry to the consideration of such normal modes was treated in Section 1.4.

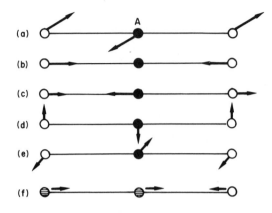

Fig. 3. Vibrational modes of CO_2 and N_2O molecules; the open circles denote O atoms, the filled ones denote C atoms, and the shaded ones denote N atoms. (a) A general vibration of a CO_2 molecule. (b)–(e) Normal modes of the CO_2 molecule. (f) A possible normal mode of a N_2O molecule.

3 to its rotation as a whole. For a linear molecule, however, rotation about the molecular axis does not involve the motion of any atom, and so only two degrees of freedom are associated with the rotation of the molecule as a whole. Thus, a linear molecule has $3n - 5$ independent modes of internal vibration, so that our CO_2 molecule has four such modes.

If we look for modes of vibration having a specific symmetry, we find one mode that is unchanged by all the symmetry operations, and this invariant mode is shown in Fig. 3b. Another mode, shown in Fig. 3c, is unaltered by rotations about the molecular axis but changes sign under the reflection σ_v. Finally, there are two independent modes, shown in Fig. 3d and e, which are unaltered by σ_v but changed by rotations about the molecular axis: since a rotation of 90° about this axis converts one into the other, these two modes must have the same vibration frequency. A general vibration of the molecule can be expressed as the sum of these four modes, which we identify as the molecule's normal modes.

We now consider possible interactions of these modes with an electric field **E**. Such an interaction can only occur if the electric dipole moment produced by the vibration has an energy in the applied field that differs from zero. This energy is, of course, a scalar quantity and so cannot be altered by any of the system's symmetry operations. In order to determine its symmetry properties, we note that it is just the sum of the scalar products of **E** with the displacement of each atom, multiplied by the atom's charge. A field perpendicular to the molecular axis has a nonzero scalar product with a mode such as that shown in Fig. 3d or e; since the field and the vibration mode are both unchanged by σ_v, while their scalar product is not changed by a rotation about this axis, an interaction can occur between these modes and an electric field parallel to them. A field parallel to the molecular axis can have nonzero scalar products with the modes shown in Fig. 3b or c. Neither this field nor these modes are affected by rotations about the molecular axis. However, while the field and the mode in Fig. 3c are both reversed by the reflection σ_v, the mode in Fig. 3b is not changed by this interaction. Hence, the interaction energy of the field with the mode in Fig. 3c is unaffected by any of the symmetry operations, so that such an interaction can occur. However, its interaction energy with the mode in Fig. 3b must be zero, as otherwise its sign would be reversed by σ_v; thus, the mode in Fig. 3b cannot interact with an external electric field.

We note one further point about this problem. If instead of CO_2 we consider the molecule N_2O, shown in Fig. 3f, which is also linear but does not possess a center of symmetry, the modes of Figs. 3b and c no longer have different symmetry, as σ_v is not a symmetry operation for N_2O. In fact, these two need not be normal modes, and in general the normal modes will be linear combinations of them, with either of which an electric field can interact.

In Section 3.3, we consider the application of symmetry considerations to the microscopic properties of quantum-mechanical systems. The two types of results that we have just discussed correspond to the classification of the system's eigenfunctions by their symmetry and the derivation of selection rules.

Symmetry of Crystal Structures

Since some of the main applications of symmetry are associated with crystals, we now discuss the main features of crystal symmetry. For crystals, it is natural to distinguish between the point symmetry, i.e., symmetry with respect to various rotations and reflections when in general at least one point remains fixed, and the translational symmetry, which we will consider first.

Let us choose three noncoplanar basic lattice vectors $\mathbf{a}_1, \mathbf{a}_2$, and $\mathbf{a}_3$. Then, if we start at a given lattice site, translations through the vectors

$$\mathbf{n} = n_1 \mathbf{a}_1 + n_2 \mathbf{a}_2 + n_3 \mathbf{a}_3, \tag{3.19}$$

where n_1, n_2, n_3 are arbitrary integers (positive, negative, or zero), bring us to a set of different lattice sites that together constitute a so-called Bravais lattice. In the simplest case, such a procedure leads us to all the sites of the lattice. In more complicated cases, however, a lattice may consist of several intersecting Bravais lattices. In such a case, in order to reach all the lattice sites we must start from a few sites (whose number equals that of the intersecting Bravais lattices), and make the set of translations **n** described by Eq. (19). Thus the simplest classification of crystal lattices is in terms of a figure bounded by the vectors a_1, a_2, and a_3, which is known as a unit cell. We start from the least symmetric system, the triclinic, for which the only symmetry elements are the identity and inversion, so that the lengths and relative orientations of a_1, a_2, and a_3 are completely arbitrary. All the other crystalline systems can readily be visualized with the help of the geometric figures, listed in Table 1, whose symmetry is contained in the corresponding crystal system. In this table, we also list the number of independent parameters (lengths of the vectors a_1, a_2, a_3 and angles between them) that determine a particular system.

We now turn to the point symmetry of crystal lattices, i.e., the symmetry between different directions in a lattice which transform into one another under the reflections and rotations compatible with the symmetry of a given lattice. The simple topological requirement that two coplanar vectors suffice to reach all the lattice sites in a given plane restricts the possible axes of symmetry for a crystal lattice to twofold, threefold, fourfold, and sixfold axes, denoted, respectively, by C_2, C_3, C_4, and C_6. If one adds to an axis of symmetry of order n an axis of second order perpendicular to it, another $n - 1$ such axes in the horizontal plane are added by the rotations of C_n; this sort of symmetry axis is known as D_n. Planes of reflection can go through the symmetry axis (vertical reflection planes, denoted by σ_v), be perpendicular to it (horizontal planes σ_h), or go through the symmetry axis and bisect the angle between adjacent horizontal symmetry axes (diagonal planes σ_d). Finally, the combination of a rotation around a C_n axis, $n = 2, 4,$ or 6, and

Table 1

Bravais Lattices

Crystal system	Characteristic geometric figure	Number of parameters	Symmetry of Bravais lattice	Number of crystal classes
Triclinic	—	6	C_i	2
Monoclinic	Straight parallelepiped	4	C_{2h}	3
Orthorhombic	Rectangular parallelepiped	3	D_{2h}	3
Tetragonal	Straight prism with square base	2	D_{4h}	7
Trigonal or Rhombohedral	Rhombohedron	2	D_{3d}	5
Hexagonal	Hexagonal prism	2	D_{6h}	7
Cubic	Cube	1	O_h	5

reflection in a plane perpendicular to the axis gives rise to a new symmetry element, namely an axis of rotatory reflection S_n.

The different sets of symmetry axes and planes are called crystal classes. As mentioned previously, a crystal can be composed of several intersecting Bravais lattices. Such a process can only decrease the symmetry of the single Bravais lattice that determines the crystal system. As a result, each crystal system contains not only the crystal class appropriate to its Bravais lattice, as listed in column 4 of Table 1, but also a number of other classes of lower symmetry.† The total number of classes in each system is listed in the last column of Table 1. A detailed description of all the 32 crystal classes may be found in textbooks on crystallography and also in those on symmetry and group theory. Each class has a different set of point symmetry operations, which is called its point group.

Two further general comments about crystal symmetry are relevant to this introduction. First, we note that a crystal lattice possesses both point and translational symmetry. Hence, for instance, the existence of one axis (or plane) of symmetry leads, because of the translational symmetry, to the appearance of an infinite number of such axes (or planes) displaced from each other by the vectors **n** defined in Eq. (19). Obviously, only one such axis (or plane) is needed to characterize the symmetry of the crystal. Moreover, the combination of a rotation or reflection and a translation in a direction parallel to the axis of rotation or to the reflection plane, respectively, can be symmetry elements for a crystal, even though neither operation on its own was such an element.‡ Such combined symmetry elements are known, respectively, as screw axes and glide (or glide reflection) planes. In this way, one can define all the sets of symmetry elements of a crystal lattice, and divide them into the 230 so-called space groups.

Our second comment is concerned with the magnetic symmetry of crystals. So far, we have only considered the symmetry of crystals with respects to different transformations of the spatial coordinates. However, in ferromagnetic and antiferromagnetic materials there are spontaneous magnetic moments, associated with the currents due to the orbital motion of the electrons and the electron spin. The reversal of such moments is associated with a change in the direction of time, $t \rightarrow -t$, so that time inversion must be

---

† The most symmetric class of the cubic system, for instance, O_h, contains all the symmetry elements of the cube, namely, three fourfold axes C_4 passing through the centers of opposite faces, four threefold axes C_3 along the cube's body diagonals, six twofold axes C_2 passing through the centers of opposite edges, plus a center of inversion symmetry at the center of the cube. The four classes of lower symmetry in the cubic system contain only some of these symmetry elements.

‡ Just as S_n can be a symmetry element in a point group even when the C_n and reflection of which it is composed are not such elements.

introduced as an extra possible symmetry element for magnetic crystals. This leads to additional symmetry classes and space groups. A more detailed description of space groups and of magnetic symmetry will be presented in Sections 3.3 and 3.4, respectively.

Symmetry of the Properties of Crystals

One major reason for the importance of crystal symmetry and its appropriate classification is the connection between the physical properties of crystals and their symmetry. According to Neumann's principle, every physical property of a crystal must possess at least the symmetry of the crystal's point group. For instance, the anisotropy of the coefficient of sound absorption in a crystal with a symmetry axis C_4 must reflect this symmetry property. Thus, this coefficient as a function of angle, can have the form shown by the solid line in Fig. 4, but not that shown by the dotted line which is associated with a threefold axis. This simple example shows the importance of symmetry requirements in the planning and analysis of experiments.

A systematic description of the physical properties of crystals, and hence of the restrictions on them imposed by the crystal's symmetry, is most readily performed by means of tensors.† The crystal symmetry can be used to reduce the number of independent nonzero elements in the tensor describing a given property of the crystal. As an example of this, let us consider the properties of a dielectric crystal subject to stresses and external electric fields. The series expansion of the free energy Φ of such a crystal, regarded as a continuous medium, involves scalar combinations of the relevant vector and tensor quantities. Its first few terms have the form

$$\Phi = \Phi_0 - (1/4\pi)E_j D_j^0 - (1/8\pi)\varepsilon_{jk}E_j E_k - \gamma_{j,kl}E_j\sigma_{kl} - \tfrac{1}{2}\mu_{jk,lm}\sigma_{jk}\sigma_{lm}.\ddagger$$

$$(3.20)$$

Here, $\mathbf{E}$ is the electric field in the dielectric (with components E_i), $\mathbf{D}$ the electric displacement, whose value is $\mathbf{D}^0$ in the absence of an external electric field, and σ_{jk} is the stress tensor of the crystal, i.e., the element σ_{jk} is the component in direction k of the force per unit area acting on a plane perpendicular to the j direction. The properties of the crystal are characterized by the dielectric tensor ε_{jk}, the piezoelectric tensor γ_{jkl}, and the elastic tensor μ_{jklm}, which in the absence of any symmetry considerations would contain $3^2 = 9$, $3^3 = 27$, and $3^4 = 81$ independent elements, respectively.

---

† For a formal definition of tensors, see Section 3.4, Eqs. (143) and (144).

‡ Note that the summation convention, i.e., summation over pairs of identical indices, is assumed again. The numerical factors in (20) are introduced for convenience in cgs Gaussian units.

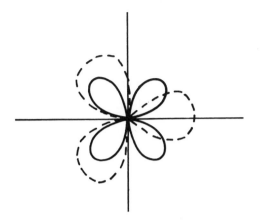

Fig. 4. Schematic forms of the coefficient of sound absorption for a crystal with a four-fold (continuous line) or threefold (broken line) symmetry axis. .

Before considering the effects of crystal symmetry, let us examine the intrinsic symmetries of these tensors. It follows from the form of the last term in (20) that $\mu_{jk,lm} = \mu_{lm,jk}$. The stress tensor is symmetric, $\sigma_{jk} = \sigma_{kj}$, because the crystal must be in mechanical equilibrium, and so $\mu_{kj,lm} = \mu_{jk,lm}$, for instance, while $\gamma_{j,kl} = \gamma_{j,lk}$. The dielectric tensor ε_{jk} is also symmetric, as we will now show. Using the well-known thermodynamic equality

$$D_j = -4\pi \, \partial\Phi/\partial E_j, \tag{3.21}$$

we find from Eq. (20) that

$$D_j = D_j^0 + \varepsilon_{jk} E_k + 4\pi\gamma_{jkl}\sigma_{kl}. \tag{3.22}$$

Hence,

$$\varepsilon_{jk} = \partial D_j/\partial E_k = -4\pi \, \partial^2\Phi/\partial E_j \partial E_k, \tag{3.23}$$

and since the second partial derivative is independent of the order of differentiation, $\partial^2\Phi/\partial E_j \, \partial E_k = \partial^2\Phi/\partial E_k \, \partial E_j$, we see that $\varepsilon_{jk} = \varepsilon_{kj}$. As a result of these intrinsic symmetries, the maximum number of independent elements in the tensors ε_{jk}, $\gamma_{j,kl}$, and $\mu_{jk,lm}$ is reduced to 6, 18, and 21, respectively.

We now turn to the effect on the vector $\mathbf{D}^0$ and the tensors ε_{jk} and γ_{jkl} of crystalline symmetry, without making use of formal group theory. Let us start with its effect on the vector $\mathbf{D}^0$ in Eq. (20). The presence of $\mathbf{D}^0$ means that the crystal is spontaneously polarized, even in the absence of an external electric field, an effect known as pyroelectricity. This pyroelectricity is not possible for every crystal symmetry. If, for instance, the crystal possesses inversion symmetry, so that every pair of opposite directions is equivalent, there can be no such vector $\mathbf{D}^0$ which marks out a preferred direction. Such a direction can only exist in crystal classes possessing a single principal axis of symmetry together with reflection planes σ_v that pass through this axis, or in classes containing no more than the identity and a single mirror plane. A simple check shows that, of the 32 crystal classes, only ten can be pyro-electric.

Let us now examine, for some typical cases, the restrictions imposed on the components of ε_{jk} and $\gamma_{j,kl}$ by the symmetry of the crystal. Consider, first of all, a crystal having a second-order axis of symmetry C_2, which we take to be the z axis. A rotation of 180° about this axis transforms the coordinates as follows: $x \rightarrow -x$, $y \rightarrow -y$, $z \rightarrow z$. Now, the components of a tensor transform according to the product of the coordinates that specify the component. Thus, for instance, in ε_{jk}, $\varepsilon_{13} = \varepsilon_{xz}$ transforms like xz, so that under this rotation $\varepsilon_{13} \rightarrow -\varepsilon_{13}$ and similarly $\varepsilon_{23} \rightarrow -\varepsilon_{23}$. However, the properties of the crystal, and hence the tensor describing them, cannot be altered by any of the crystal's symmetry operations. Thus, in our case, $\varepsilon_{13} = -\varepsilon_{13}$ and $\varepsilon_{23} = -\varepsilon_{23}$, i.e., $\varepsilon_{13} = \varepsilon_{23} = 0$. The effect of other symmetry elements can be examined similarly. It emerges that for all classes belonging to the same crystal system (see Table 1) the number of independent components of ε_{jk} is the same. This result is in contrast to that for tensors of higher rank, where the number of independent components depends, in general, on the symmetry class of the crystal, and not just on the system. A symmetric second rank tensor such as ε_{jk} has three independent components for crystals belonging to the orthorhombic system, for instance, and two for those belonging to the tetragonal, rhombohedral, and hexagonal systems, but only one for cubic crystals. For these latter crystals, one can write $\varepsilon_{jk} = \varepsilon\delta_{jk}$, so that their dielectric tensor is the same as for an isotropic body.†

We now turn to the tensor $\gamma_{j,kl}$ and continue to consider a crystal having just a single twofold axis C_2. Since a rotation of 180° about this axis reverses the signs of both x and y, any tensor component for which x and y together appear an odd number of times must be zero. Thus for instance, both $\gamma_{1,11}$ (corresponding to $\gamma_{x,xx}$) and $\gamma_{2,11}$ (corresponding to $\gamma_{y,xx}$) vanish in such a crystal. As a result, of the 18 possible nonzero independent components of $\gamma_{j,kl}$, this single symmetry operation causes ten to vanish, and the only independent components that can differ from zero are $\gamma_{1,13}$,

---

† Any symmetric second-order tensor can be diagonalized by an appropriate choice of the coordinate axes. A result of tensor algebra is that such tensors can be represented geometrically by an ellipsoid, the lengths of whose axes are proportional to the principal values ε_{11}, ε_{22}, and ε_{33} of the tensor. The symmetry of this ellipsoid corresponds to that of the crystal, and it can degenerate into an ellipsoid of rotation ($\varepsilon_{11} = \varepsilon_{22} \neq \varepsilon_{33}$) or a sphere ($\varepsilon_{11} = \varepsilon_{22} = \varepsilon_{33}$).

$\gamma_{1,23}$, $\gamma_{2,13}$, $\gamma_{2,23}$, $\gamma_{3,11}$, $\gamma_{3,12}$, $\gamma_{3,22}$, and $\gamma_{3,33}$.[†] The influence of other symmetry elements on the number of independent components of the piezoelectric tensor $\gamma_{j,kl}$ can be found similarly. In this way, it is possible to find the number of and identify the independent nonzero components of tensors of any rank for all symmetry classes. This sort of prediction, which is quite independent of the microscopic model used to account for the property being considered, is one very important type of application of the theory of symmetry to the equilibrium properties of crystals. Some other examples, including ones involving magnetic fields and axial tensors, will be presented in Section 3.4.

*The Symmetry of Kinetic Coefficients—Onsager's Principle

The principle of the symmetry of the kinetic coefficients, which is known as Onsager's principle, forms the basis of the thermodynamics of irreversible processes. In order to describe it, we consider a nonequilibrium state of a closed system, described by the set of quantities $x_1, \ldots, x_n$. For each x_j, we define a conjugate parameter X_j in terms of the partial derivative of the system's entropy S with respect to x_j,

$$X_j = -\partial S/\partial x_j. \tag{3.24}$$

If the system were left on its own, it would gradually approach equilibrium, with each parameter x_j approaching its equilibrium value with the velocity

$$\dot{x}_j \equiv dx_j/dt. \tag{3.25}$$

In equilibrium, the entropy is a maximum, so that all its first derivatives, and so the parameters X_j, vanish. Similarly, in equilibrium each $\dot{x}_j$ is zero. Hence, in a linear approximation we can expect that

$$\dot{x}_j = -\gamma_{jk}X_k, \tag{3.26}$$

where the constants γ_{jk} are called Onsager's kinetic coefficients.

We now turn to the proof that these coefficients are symmetric. Let the value of x_j at the time t be denoted by $x_j(t)$. The state of the system at an arbitrary time $t + \tau$ is not determined solely by the parameter $x_j(t)$. Hence, we can only discuss the correlations between x_k at such a time, $x_k(t + \tau)$, and $x_j(t)$, averaged over all possible states with a given $x_j(t)$. Thus, we define the correlation function

$$f_{jk}(\tau) = \overline{x_j(t)x_k(t + \tau)} = \overline{x_j(0)x_k(\tau)}. \tag{3.27}$$

[†] Since, as we showed above, $\gamma_{j,lk} = \gamma_{j,kl}$, the other nonzero components of $\gamma_{j,kl}$ will be equal to one of these eight.

In this definition, we have used the obvious fact that a correlation function cannot depend on the choice of the origin of time t, but only on the time difference τ. Onsager's main idea was to apply the principle of microscopic reversibility to these correlation functions. According to this principle, the equation of state is invariant with respect to a change in the sign of the time, $t \to -t$, so that $f_{jk}(\tau) = f_{jk}(-\tau)$. Thus

$$\overline{x_j(0)x_k(\tau)} = \overline{x_j(0)x_k(-\tau)} = \overline{x_j(\tau)x_k(0)}, \tag{3.28}$$

where the last equality holds because the correlation function is unaltered by a shift through τ of the origin of time. On subtracting $\overline{x_j(0)x_k(0)}$ from both sides of Eq. (28), dividing by τ, and taking the limit $\tau \to 0$, we find that

$$\overline{x_j(0)\dot{x}_k(0)} = \overline{\dot{x}_j(0)x_k(0)}. \tag{3.29}$$

Thus, on substituting from Eq. (26) for $\dot{x}$, we obtain the equation

$$-\gamma_{kl}\overline{x_j(0)X_l(0)} = -\gamma_{jm}\overline{X_m(0)x_k(0)}. \tag{3.30}$$

In order to evaluate the correlation functions that appear in Eq. (30), we must integrate over all the variables x_j, giving to each set of these variables its appropriate statistical weight, $\exp[(S - S_0)/k_B]$, corresponding to its degeneracy. Thus, in general,

$$\overline{(x_p(0)X_q(0))} = \int x_p X_q \exp\left(\frac{S - S_0}{k_B}\right) dx_1\, dx_2 \cdots dx_n$$

$$= -k_B \int x_p \frac{\partial}{\partial x_q} \exp\left(\frac{S - S_0}{k_B}\right) dx_1 \cdots dx_n \tag{3.31a}$$

$$= k_B \int \frac{\partial x_p}{\partial x_q} \exp\left(\frac{S - S_0}{k_B}\right) dx_1 \cdots dx_n = \delta_{p,q}. \tag{3.31b}$$

Here, Eq. (24) was used to derive (31a), and we then integrated by parts with respect to x_q, while the appropriate boundary conditions and normalization associated with S lead to our final result. On substituting from (31) into (30), we find that $\gamma_{kj} = \gamma_{jk}$, i.e., that Onsager's kinetic coefficients are symmetric, a result known as Onsager's principle.

In order to use Onsager's principle in practice, we must first choose the parameters x_j and X_j. To do this, we require a formula for the rate of change of the entropy in terms of these parameters and in terms of known variables. Since entropy is an extensive variable, proportional to the volume of the system, it is convenient henceforth to take the parameters x_j to represent densities. Then

$$\frac{dS}{dt} = \int \sum_j \frac{\partial S}{\partial x_j} \frac{dx_j}{dt} dv = -\sum_j \int X_j \dot{x}_j \, dv, \tag{3.32}$$

where the integration is over the volume of the system.

As our first application of the principle, let us consider the electrical conductivity. The quantity of heat evolved per unit time is determined by Joule's law,

$$\frac{dQ}{dt} = \int \mathbf{j} \cdot \mathbf{E} \, dv, \tag{3.33}$$

where $\mathbf{j}$ is the electrical current density and $\mathbf{E}$ the electric field. Now, from thermodynamics, we know that the heat evolved dQ is related to the change in entropy dS by

$$dQ = T \, dS, \tag{3.34}$$

where T is the absolute temperature. Hence, we find that

$$\frac{dS}{dt} = \int \frac{\mathbf{j} \cdot \mathbf{E}}{T} \, dv, \tag{3.35}$$

an equation that has the form of (32) with $\dot{x}_k = j_k$ and $X_k = -E_k/T$. For an anisotropic body, the relation between j and E is given by Ohm's law,

$$j_l = \sigma_{lm} E_m, \tag{3.36a}$$

i.e.,

$$\dot{x}_l = -\sigma_{lm} T X_m. \tag{3.36b}$$

On comparing this equation with (26), we see that $\gamma_{lm} = T\sigma_{lm}$. Hence, Onsager's principle leads to the symmetry of the conductivity tensor, $\sigma_{lm} = \sigma_{ml}$. Since this conductivity tensor describes a nonequilibrium property, the proof of its symmetry required more sophisticated arguments than those required for equilibrium properties (such as the dielectric tensor ε_{jk}, whose symmetry we derived above from the free energy).

Our second example relates to thermoelectric phenomena, when a conductor having a nonuniform temperature distribution is placed in a static external electric field $E = -\nabla\phi$. The temperature gradient ∇T will lead to a net flow of electrons, and so will produce an electric current even if $E = 0$. Thus, in general,

$$\mathbf{j} = \sigma \mathbf{E} - \alpha \, \nabla T. \tag{3.37}$$

For simplicity, we consider an isotropic system (or one of cubic symmetry—see above), for which the conductivity $\sigma_{lm} = \sigma\delta_{l,m}$, and a system for which **E**, if nonzero, is parallel to ∇T, so that Eq. (37) can be replaced by the corresponding scalar equation. The phenomenological coefficient α characterizes a thermoelectric phenomenon, i.e., the value of the current in the absence of an electric field. In addition to the flux of electrical charge represented by the current density **j**, there is also a flux of energy, of density **q**. This is determined by the energy $j\phi$ transferred convectively by the electrons,[†] the heat conducted as a result of the temperative gradient ∇T, and also by a cross-effect associated with the electric field **E**, similar to the $\alpha\,\nabla T$ term in Eq. (37). Thus we write, denoting by $\tilde{\nabla}T$, the magnitude of DT,

$$q - j\phi = -\kappa\tilde{\nabla}T + \beta E. \qquad (3.38)$$

As we now proceed to show, Onsager's principle gives us a relation between the coefficients α and β appearing in the last two equations. Because of the conservation of energy, the amount of heat evolved per unit time dQ/dt is $\int -\tilde{\nabla}\cdot\mathbf{q}\,dv$, and so, from (34),

$$\frac{dS}{dt} = -\int \frac{\nabla\cdot\mathbf{q}}{T}\,dv \equiv -\int\frac{1}{T}\nabla\cdot(\mathbf{q} - \phi\mathbf{j})\,dv - \int\frac{1}{T}\nabla\cdot(\phi\mathbf{j})\,dv. \qquad (3.39a)$$

On transforming the first of these integrals by partial integration, and using in the second the fact that $\nabla\cdot\mathbf{j} = 0$ as the current is constant, we readily find that for our system

$$\frac{dS}{dt} = -\int\frac{q - \phi j}{T^2}\tilde{\nabla}T\,dv + \int\frac{Ej}{T}\,dv. \qquad (3.39b)$$

On comparing this with Eq. (32), we see that we can identify

$$\dot{x}_1 = j, \qquad X_1 = -E/T, \qquad \dot{x}_2 = q - \phi j, \qquad X_2 = \tilde{\nabla}T/T^2. \quad (3.40)$$

In terms of these variables, our Eqs. (37) and (38) become

$$\dot{x}_1 = -\sigma T X_1 - \alpha T^2 X_2, \qquad \dot{x}_2 = -\beta T X_1 - \kappa T^2 X_2. \qquad (3.41)$$

In accordance with Onsager's principle, the cross-coefficients in this equation are equal, i.e., $\alpha T = \beta$. Thus, the use of Onsager's principle has reduced by one the number of unknown phenomenological coefficients, thereby facilitating both their experimental measurement and their theoretical calculation.

Two further points are worth noting with regard to Onsager's relations. First, their derivation was based on the invariance of mechanical equations

† We define ϕ to include also the Fermi energy of the electrons in the absence of the electric field.

with respect to the operation of time reversal, $t \to -t$. However, if a magnetic field $\mathbf{H}$ or angular velocity $\boldsymbol{\Omega}$, for instance, are present among the parameters x_j that characterize the state of the system, the equations of motion will only remain invariant under time inversion if the signs of $\boldsymbol{\Omega}$ and $\mathbf{H}$ are also changed.† Hence, in this case the principle of symmetry of the kinetic coefficients takes the form

$$\gamma_{jk}(\mathbf{H}, \boldsymbol{\Omega}) = \gamma_{kj}(-\mathbf{H}, -\boldsymbol{\Omega}). \tag{3.42}$$

Second, the derivation of Onsager's relations also contains the implicit assumption that the quantities $x_j(t)$ are unchanged by time reversal. If both x_j and x_k change sign under time reversal, the result $\gamma_{jk} = \gamma_{kj}$ will be unchanged. However, if only one of them changes sign when $t \to -t$, Onsager's relations will obviously be changed to

$$\gamma_{jk} = -\gamma_{kj}. \tag{3.43}$$

Order–Disorder Phase Transitions

Finally, as an example of the use of symmetry to construct and provide a basis for a physical theory, we consider the theory of order–disorder phase transitions, in which a continuous change in the state of the body is accompanied by a discontinuous change in its symmetry. In Section 2.5 we discussed scaling theory, which is the modern version of order–disorder phase transitions and is based on dimensional analysis. Here, we start with a phenomenological version of the theory, proposed by Landau in 1937, and then consider his microscopical justification of the theory on the basis of symmetry arguments, which leads to important general predictions.

The thermodynamic state of a homogeneous system can be described by the Helmholtz free energy Φ, which depends on the pressure p and temperature T, i.e., $\Phi = \Phi(p, T)$. In general, the number of independent variables describing the thermodynamic state of a system equals the number of its degrees of freedom, which for the homogeneous one-component system under consideration is equal to two. An additional parameter η is introduced to describe the phase transition. This parameter serve to distinguish between the two phases, and we define it to be zero for one phase and nonzero for the other. Such a parameter cannot affect the number of degrees of thermodynamic freedom, and so it cannot be independent and must be a function of the thermodynamic variables describing the system, i.e., of p and T in our case. Since, in equilibrium, a system's free energy is minimal, we require

---

† This can be seen from the fact that only such a change leaves unaltered the Coriolis and Lorentz forces, which are proportional to $\mathbf{v} \times \boldsymbol{\Omega}$ and $\mathbf{v} \times \mathbf{H}$, respectively, when t is replaced by $-t$.

Φ, when regarded as a function of η, to be a minimum, and this determines Φ as a function of p and T. Thus, η is determined by the pair of conditions

$$\partial\Phi/\partial\eta = 0, \qquad \partial^2\Phi/\partial\eta^2 > 0. \tag{3.44}$$

Since η is zero on one side of the phase transition point, we expect it to remain small in the immediate vicinity of such a point, so that we can usefully expand Φ as a series in the powers of η,

$$\Phi = \Phi_0 + A(p, T)\eta^2 + B(p, T)\eta^3 + C(p, T)\eta^4 + \cdots. \tag{3.45}$$

On one side of the transition point, which we call "above" it, the minimum value of Φ is for $\eta = 0$. This requires both that Φ does not contain a term linear in η and that $A(p, T)$ be positive in this region. In order for a minimum to occur with nonzero η, i.e., for the transition to the "lower" phase, the function $A(p, T)$ must change sign and become negative. Thus, phase transition points are determined by the equation

$$A(p, T) = 0. \tag{3.46}$$

However, this condition is not sufficient, since for Φ to have a minimum at $\eta = 0$ its series expansion must start with an even power of η. Thus, at a phase transition point we also require that

$$B(p, T) = 0. \tag{3.47}$$

Equations (46) and (47) will in general determine a single transition point, which is an isolated point on the phase diagram. However, situations can arise in which the term of third order in η in Eq. (45) is identically zero, so that Eq. (47) is satisfied everywhere. In this case, which is the more interesting one, we obtain a line of transition points lying on the curve $p = p(T)$ determined by Eq. (46). We shall return later to the conditions for terms in η^3 not to appear in Eq. (45), conditions that are associated with the symmetry of the system.

As a simple example, we consider first the ferromagnetic phase transition of a cubic crystal. For this, we choose as our parameter η of Eq. (45) the spontaneous magnetization $\mathbf{M}$. Since Φ is a scalar quantity, it cannot involve odd powers of the vector $\mathbf{M}$, and so the third-order term is identically zero. In this case, the requirements of Eq. (44), applied to the series (45) with $B \equiv 0$ and $\eta^2 = M^2$, become

$$2AM + 4CM^3 + \cdots = 0, \tag{3.48}$$

$$2A + 12CM^2 + \cdots > 0. \tag{3.49}$$

On retaining only these first two terms in each of these equations, we see that Eq. (48) has two solutions, namely, $M = 0$ for the "upper" phase, and $M = (-A/2C)^{1/2}$ for the "lower" phase. The inequality (49) will then only be satisfied if the function $A(p, T)$ has opposite signs on the two sides of the transition point, and so vanishes at this point, as we noted earlier. In order to obtain the temperature dependence of M for a given value of the pressure p, let us expand $A(p, T)$ into a power series in the region of the transition temperature T_c appropriate to this pressure. Thus, we write

$$A(T) = a(T - T_c) + \cdots, \tag{3.50}$$

and hence we find that in the "lower" phase†

$$M = \sqrt{a(T_c - T)/2C}. \tag{3.51}$$

Equations (50) and (51), in conjunction with (45), enable us to find the temperature dependence of all the thermodynamic variables in the region of the transition point T_c.‡

The above theory is a purely phenomenological one, and so leaves us with some feeling of dissatisfaction at the arbitrary introduction of the parameter η and the series (45) for the free energy Φ. We shall, now show, for the example of the transition from a liquid to a crystalline solid, how the series (45) can arise. The subsequent generalization of this treatment will give us an insight into the general approach to order–disorder phase transitions in terms of the theory of symmetry.

Let the constant ρ_0 be the density of the liquid, which we refer to as the "upper" phase. At the phase transition point (which is just the melting temperature at a given pressure), the density starts to change and becomes $\rho_0 + \delta\rho$, where $\delta\rho$ has the symmetry of the crystalline "lower" phase. We expand $\delta\rho$ in a series of plane waves

$$\delta\rho = \sum_{\mathbf{g}} a_{\mathbf{g}} \exp(i\mathbf{g} \cdot \mathbf{r}), \qquad a_{-\mathbf{g}} = a_{\mathbf{g}}^*. \tag{3.52}$$

The free energy Φ of the crystal depends on its density, and so can be expanded near the transition point as a series in $\delta\rho$, or equivalently in the coefficients $a_{\mathbf{g}}$. The terms in this expansion are of the form $a_{\mathbf{g}_1} a_{\mathbf{g}_2} \cdots a_{\mathbf{g}_n} \exp(i(\mathbf{g}_1 + \mathbf{g}_2 + \cdots + \mathbf{g}_n) \cdot \mathbf{r})$. However, the free energy must be invariant under a translation of the origin of the system of coordinates, $\mathbf{r} \to \mathbf{r} + \mathbf{R}$, and so the expansion for Φ can only contain terms for which $\mathbf{g}_1 + \mathbf{g}_2 + \cdots + \mathbf{g}_n = 0$.

The above considerations show that the only first-order terms in the expansion have $\mathbf{g} = 0$, i.e., these are only constants, just as we postulated

---

† In the present approximation, the coefficient $C(T)$ can be put equal to its value at T_c and so treated as a constant.

‡ As is well known, the results of the theory presented here are not consistent with the experimental data in the immediate vicinity of the transition point.

for Eq. (45). Terms of the second order contain only the products $a_{\mathbf{g}} a_{-\mathbf{g}}$, which is just $|a_{\mathbf{g}}|^2$ in view of (52). Thus, to second order

$$\Phi = \Phi_0 + \sum_{\mathbf{g}} A_{\mathbf{g}} |a_{\mathbf{g}}|^2. \tag{3.53}$$

In order that $\delta\rho$ be nonzero in the "lower" phase, at least one of the $A_{\mathbf{g}}$, $A_{\mathbf{g}_0}$ say, must become nonzero. The point at which this $A_{\mathbf{g}_0}$ vanishes is a transition point, and "below" it a density fluctuation $\delta\rho$ arises which is represented by a plane wave of wavelength $2\pi/|g_0|$. All the other $A_{\mathbf{g}}$ in Eq. (53) will remain zero, so that near the transition point Eq. (53) becomes

$$\Phi = \Phi_0 + A_{g_0} \sum_{|\mathbf{g}| = g_0} |a_{\mathbf{g}}|^2. \tag{3.54}$$

Here A is a function only of $g_0 = |\mathbf{g}_0|$ since a liquid is isotropic.

The third-order terms in Φ will be of the form $B_{\mathbf{g}} a_{\mathbf{g}_1} a_{\mathbf{g}_2} a_{\mathbf{g}_3}$ where $\mathbf{g}_1 + \mathbf{g}_2 + \mathbf{g}_3 = 0$. However, near the transition point only those $a_{\mathbf{g}}$ with $|\mathbf{g}| = g_0$ are nonzero. Thus, to third order in the $a_{\mathbf{g}}$, we can write

$$\Phi = \Phi_0 + A_{g_0} \sum_{|\mathbf{g}| = g_0} |a_{\mathbf{g}}|^2 + B_{g_0} \sum a_{\mathbf{g}_1} a_{\mathbf{g}_2} a_{\mathbf{g}_3}, \tag{3.55}$$

where in the third term $|\mathbf{g}_j| = g_0$, the vectors $\mathbf{g}_1$, $\mathbf{g}_2$, and $\mathbf{g}_3$ form an equilateral triangle, and the summation is over the different orientations of this triangle in space. A comparison of Eq. (55) with (45) shows that for the phase transition from liquid to crystal we have managed to construct an expansion of the free energy Φ in terms of an order parameter, $a_{\mathbf{g}}$ in this case, which contains terms of the third order. According to our previous discussion, this means that a transition from a liquid phase to a crystalline solid can only occur at an isolated point on the phase diagram. Thus, the transition from a liquid to a crystalline solid is not of the order–disorder type.

A simple generalization of the method used in this example enables us to see how the theory of symmetry can provide, in general, a basis for the phenomenological theory and lead to predictions of both theoretical and practical importance. Just as for the transition from a liquid to a crystalline solid, it is possible in general to represent the density (or some other property that distinguishes between the two phases) in the "lower" phase as a linear combination of some function $f_1, f_2, \ldots$ which are transformed into one another by the symmetry elements of the "upper" phase. Thus, we write

$$\delta\rho = \sum_j c_j f_j \tag{3.56}$$

as a generalization of Eq. (52). In that case, the functions f_j were plane waves $\exp(i\mathbf{g} \cdot \mathbf{r})$, which transformed into each other (for each $|\mathbf{g}| = \text{const}$) under the symmetry elements of the isotropic system, where the density was ρ_0, a constant. In general, one can choose the functions f_j to belong to a series of sets $f_j^{(n)}$ (the bases of so-called irreducible representations of the symmetry group),† each of which transforms into itself. Thus, we write

$$\delta\rho = \sum_j \sum_n c_j^{(n)} f_j^{(n)} \tag{3.57}$$

The free energy Φ depends on $\rho_0 + \delta\rho$, so that near the transition point we can expand Φ as a series in the powers of $c_j^{(n)}$. The functions $f_j^{(n)}$ are just basis sets, so that it is natural to consider Φ as a function of the $c_j^{(n)}$, in analogy with our use of a_g in Eqs. (52)–(55). The expansion of Φ must contain only sets of terms that are invariant under the symmetry elements of the "upper" phase, just as in the derivation of Eqs. (53)–(55). There is a general theorem that no linear invariant can be constructed from quantities that transform according to an irreducible representation. Thus, Φ cannot contain a term linear in the order parameter, in agreement with Eq. (45). The possibility of constructing invariants of the third order is, as we have seen, the crucial question that determines whether the phase diagram contains a line along which order–disorder phase transitions will occur. Hence, we can investigate theoretically whether such phase transitions are possible by analyzing all the different pairs of symmetries for the "lower" and "upper" phases. This is a clear example of the heuristic power of the theory of symmetry for the analysis of physical phenomena. In Section 3.3 we shall return to this subject in more detail.

3.2 Conservation Laws in Quantum Mechanics

We saw in Section 3.1 that the laws of conservation of energy, momentum, and angular momentum in classical mechanics are an immediate consequence of the homogeneity of time and space and of the isotropy of space. These and the other classical conservation laws considered there can also be formulated and derived within the framework of quantum mechanics. In addition, there are new conservation laws in quantum mechanics, associated with features such as parity and the indistinguishability principle. In this section, we consider all these conservation laws, and also examine time-reversal symmetry and some of its consequences.

Quantum-Mechanical Formulation of Conservation Laws

In nonrelativistic quantum mechanics, the state of a physical system is determined by its wave function $\psi(\mathbf{r}, t)$, where $\mathbf{r}$ denotes the set of positions

† These are discussed in Section 3.3.

$\mathbf{r}_k$ of all the particles in the system. The variation of this function with time is determined by Schrödinger's equation

$$i\hbar \, \partial\psi/\partial t = \hat{H}\psi, \tag{3.58}$$

where $\hat{H}$ is the system's Hamiltonian operator. In contrast to classical mechanics, only a restricted number of physical quantities $A, B, \ldots$ have definite values $a, b, \ldots$ for each state ψ. Each of these values is the solution of the appropriate eigenvalue equation, of the form

$$\hat{A}\psi = a\psi, \tag{3.59}$$

where $\hat{A}$ is the operator associated with the physical quantity A. If the wave function ψ is normalized, so that

$$\int \psi^*(\mathbf{r})\psi(\mathbf{r}) \, dv = 1, \tag{3.60}$$

we can rewrite Eq. (59)

$$a = \int \psi^*(\mathbf{r})\hat{A}\psi(\mathbf{r}) \, dv. \tag{3.61}$$

For instance, the possible values of a system's energy are the eigenvalues of its Hamiltonian operator $\hat{H}$, i.e., the numbers E_n such that

$$\hat{H}\psi_n = E_n\psi_n; \qquad E_n = \int \psi_n^*(\mathbf{r})\hat{H}\psi_n(\mathbf{r}) \, dv. \tag{3.62}$$

Here the subscript n is associated with the nth state, which has wave function ψ_n and energy E_n.

On differentiating Eq. (61) with respect to time, we find that

$$\frac{da}{dt} = \int \left[\frac{\partial\psi^*}{\partial t} \hat{A}\psi + \psi^*\hat{A} \frac{\partial\psi}{\partial t} \right] dv + \int \psi^*(\mathbf{r}) \frac{\partial\hat{A}}{\partial t} \psi(\mathbf{r}) \, dv. \tag{3.63}$$

Hence, on substituting from Eq. (58) and its complex conjugate for $\partial\psi/\partial t$ and $\partial\psi^*/\partial t$ and using the fact that $\hat{H}$ is a Hermitian operator, we find that

$$\frac{da}{dt} = \int \psi^*(\mathbf{r}) \left[\frac{\partial\hat{A}}{\partial t} + \frac{i}{\hbar}(\hat{H}\hat{A} - \hat{A}\hat{H}) \right] \psi(\mathbf{r}) \, dv. \tag{3.64}$$

On comparing Eqs. (64) and (61), we see that the operator corresponding to da/dt is that appearing in the integrand of Eq. (64). If the operator $\hat{A}$ does not depend explicitly on the time, so that $\partial\hat{A}/\partial t = 0$, the mean value of the observable A will remain constant in time (so that a is a constant of the motion or conserved quantity) if

$$\hat{H}\hat{A} = \hat{A}\hat{H}. \tag{3.65}$$

Thus, a quantity A that does not depend explicitly on time will be conserved if and only if its operator commutes with the Hamiltonian operator $\hat{H}$.

The Conservation of Energy, Momentum, and Angular Momentum

The simplest example of an operator $\hat{A}$ that commutes with $\hat{H}$ is $\hat{H}$ itself. Therefore, energy is a conserved quantity, with its possible values E_n given by Eq. (62), provided that $\partial\hat{H}/\partial t = 0$. This condition will be fulfilled if the system is not in a varying external field, since then all instants of time are equivalent [i.e., time is homogeneous, cf. Eq. (5)], so that the Hamiltonian operator $\hat{H}$ cannot involve the time explicitly.

We turn next to the conservation of momentum, which follows from the homogeneity of space. For a system not exposed to an external field (a closed system), all positions in space are equivalent. In other words, the Hamiltonian operator cannot be changed when the origin of coordinates is moved through an infinitesimal distance $-\delta\mathbf{r}$, or equivalently when all the particles in the system are displaced through the same distance $\delta\mathbf{r}$, so that $\mathbf{r}_k \to \mathbf{r}_k + \delta\mathbf{r}$. Such a displacement transforms the system's wave function $\psi(\mathbf{r}_1, \mathbf{r}_2, \ldots)$, to first order in $\delta\mathbf{r}$, into

$$\psi(\mathbf{r}_1 + \delta\mathbf{r}, \mathbf{r}_2 + \delta\mathbf{r}, \ldots) = \psi(\mathbf{r}_1, \mathbf{r}_2, \ldots) + \delta\mathbf{r} \cdot \sum_k \nabla_k \psi$$

$$\equiv \left(1 + \delta\mathbf{r} \cdot \sum_k \nabla_k\right)\psi(\mathbf{r}_1, \mathbf{r}_2, \ldots). \qquad (3.66)$$

Thus, the expression $(1 + \delta\mathbf{r} \cdot \Sigma_k \nabla_k)$ is the operator corresponding to a displacement of the system by $\delta\mathbf{r}$. In view of the homogeneity of space and Eq. (65), if the displacement does not depend on time, this operator must commute with $\hat{H}$, so that

$$\left(\sum_k \nabla_k\right)\hat{H} = \hat{H} \sum_k \nabla_k. \qquad (3.67)$$

The quantity whose conservation follows from the homogeneity of space is called the momentum, and so the operator ∇_k must be proportional to the momentum of the kth particle. In order to obtain the classical formula for the momentum on making the transition in the limit from quantum to classical mechanics, we define the momentum operator for the kth particle as $-i\hbar\nabla_k$, and that for the whole system as

$$\hat{\mathbf{P}} = -i\hbar \sum_k \nabla_k. \qquad (3.68)$$

From the isotropy of space, i.e., the equivalence of all directions in it, it follows that the Hamiltonian operator of a closed system cannot change when the system (or its coordinate axes) is rotated as a whole. When the system rotates through an infinitesimal angle $\delta\phi$, each point $\mathbf{r}_k$ is displaced by $\delta\mathbf{r}_k = \delta\phi \times \mathbf{r}_k$, in accordance with Eq. (7a). Hence, by analogy with equation (66), the operator corresponding to such an infinitesimal rotation is just

$$1 + \sum_k (\delta\phi \times \mathbf{r}_k) \cdot \nabla_k = 1 + \delta\phi \cdot \sum_k (\mathbf{r}_k \times \nabla_k), \tag{3.69}$$

where the equality follows from the formula for the scalar triple product $(\mathbf{a} \times \mathbf{b}) \cdot \mathbf{c} = \mathbf{a} \cdot (\mathbf{b} \times \mathbf{c})$. The invariance of the state of the system under such a rotation (when this does not depend explicitly on time) then requires, in view of Eq. (65), that

$$\sum_k (\mathbf{r}_k \times \nabla_k)\hat{H} = \hat{H} \sum_k (\mathbf{r}_k \times \nabla_k). \tag{3.70}$$

The quantity whose conservation follows from the isotropy of space is called the angular momentum, and so $\mathbf{r}_k \times \nabla_k$ must be proportional to the angular momentum of particle k. Just as for the ordinary momentum, the constant of proportionality is chosen to be $-i\hbar$, so that the angular momentum operator of a single particle $\hat{\mathbf{L}}$ is defined by

$$\hat{\mathbf{L}} = -i\hbar\mathbf{r} \times \nabla = \mathbf{r} \times \hat{\mathbf{p}}, \tag{3.71}$$

where $\hat{\mathbf{p}}$ is the momentum operator.

For a system in an external field, angular momentum is not, in general, conserved. However, the symmetry of the external field can lead to the conservation of angular momentum or some of its components, just as for classical mechanics. For instance, angular momentum is conserved in a centrally symmetric field, where all directions in space are still equivalent, while in an axially symmetric field, the projection of the angular momentum along the symmetry axis of the field is conserved.

Thus, the general symmetries of time and space lead to the conservation of energy, momentum, and angular momentum in quantum mechanics just as they did in classical mechanics.

Parity

As the first example of a conservation law that is specifically quantum mechanical and has no simple classical counterpart, we consider the conservation of parity.

Apart from translations and pure rotations in space, another operation that leaves invariant the Hamiltonian operator of a closed system is that of inversion, i.e., the simultaneous change of sign of all the spatial coordinates, so that $\mathbf{r} \to -\mathbf{r}$. This will also be a symmetry operation, of course, for a system in an external field whose potential $V(\mathbf{r})$ satisfies $V(-\mathbf{r}) = V(\mathbf{r})$. Let us denote the inversion operator by $\hat{P}$. If it leaves the Hamiltonian operator $\hat{H}$ invariant, $\hat{P}$ commutes with $\hat{H}$. Hence, from Eq. (62), if $\psi_n(\mathbf{r})$ is an eigenfunction of $\hat{H}$ then so is

$$\hat{P}\psi_n(\mathbf{r}) = \psi_n(-\mathbf{r}). \tag{3.72}$$

Incidentally, Eq. (72) for arbitrary ψ can be regarded as a definition of the inversion operator $\hat{P}$. On applying $\hat{P}$ again to Eq. (72), we find that

$$\hat{P}^2\psi_n = \psi_n. \tag{3.73}$$

Thus, the inversion operator $\hat{P}$ has just two possible eigenvalues ± 1. A state with eigenvalue $+1$, i.e., which is unchanged by inversion, is said to be of even parity, while one with eigenvalue -1, so that its wave function changes sign on inversion, is said to have odd parity. Since $\hat{P}$ commutes with $\hat{H}$, this property is conserved during time, and the state of a closed system always has a fixed parity.

The conservation of parity plays an important role in the matrix formulation of quantum mechanics, in which the main element of the theory is the matrix element a_{mn} corresponding to the transition from state n to state m under the influence of operator $\hat{A}$. This matrix element is defined by

$$a_{mn} = \int \psi_m^*(\mathbf{r})\hat{A}\psi_n(\mathbf{r}) \, dv. \tag{3.74}$$

Usually, the operator $\hat{A}$ has a definite parity; for instance, we have assumed that $\hat{H}$ is invariant under $\hat{P}$, and so is of even parity, while the electric field is of odd parity, as we show later.

For an operator of even parity, which we denote by $\hat{A}_+$, a_{mn} can differ from zero only if states m and n are of the same parity. For if they have different parity, for instance if ψ_m is odd and ψ_n even, the integrand of Eq. (74) will change sign under the inversion operation $\mathbf{r} \to -\mathbf{r}$. However, this operation cannot change the value of the integral over all space, so that in this case $a_{mn} = -a_{mn} = 0$. Thus, the matrix elements of an operator of even parity between states of opposite parity must vanish. Analogous arguments show that the matrix elements of an operator of odd parity $\hat{A}_-$ between states having the same parity must vanish. In the analysis and design of experiments, it is very important to know which matrix elements vanish, so that the corresponding transitions are forbidden. We shall return to this subject of selection rules in Section 3.3.

As an example of the application of parity, we consider a particle moving in a centrally symmetric field, such as an electron in a hydrogen atom. It is known from elementary quantum mechanics that the wave function of such a particle is of the form, in spherical polar coordinates r, θ, ϕ,

$$\psi_{nlm}(\mathbf{r}) = F_n(r)P_l^m(\cos\theta)e^{im\phi}, \tag{3.75}$$

where the functions P_l^m are the associated Legendre polynomials and $F_n(r)$ is some function of the scalar r. In these coordinates, the inversion operation is described by $r \to r$, $\theta \to \pi - \theta$, $\phi \to \pi + \phi$. Since $\cos(\pi - \theta) = -\cos\theta$, it follows from the theory of Legendre polynomials that the replacement of θ by $\pi - \theta$ multiplies $\psi(\mathbf{r})$ by $(-1)^{l-m}$, while the replacement of ϕ by $\pi + \phi$ introduces a factor $e^{im\pi} = (-1)^m$. Hence, the wave function $\psi_{nlm}(\mathbf{r})$ is multiplied by a factor $(-1)^l$ when the spatial coordinates are inverted and so has odd or even parity according as l is odd or even. Thus, an operator of odd parity, such as the electric dipole moment, will have zero matrix elements between a pair of states for both of which l is odd or for both of which l is even.

For a system of particles in a centrally symmetric field, if the mutual interaction of the particles is weak, each particle can be regarded as having a definite parity. If that of the jth particle is $(-1)^{l_j}$, then the parity of the whole system will be just $(-1)^{l_1+l_2+\cdots}$.† Some interesting conclusions about the possible decay of a system of particles follow from this result. Even if the requirements of energy conservation permit a decay, it can be forbidden because of parity conservation.

Time-Reversal Symmetry in Classical Physics

In a simple classical mechanical system there is no preferred direction of time. Since the equations of motion of classical mechanics are differential equations involving second derivatives with respect to time t, they are invariant under the time-reversal transformation $t \to -t$. This transformation changes the sign not only of t but also of all odd functions of t, such as the velocity $\mathbf{v}$. As a result, a difference arises between static and dynamic quantities, such as the charge density $\rho(\mathbf{r}, t)$ and the current density $\mathbf{j}(\mathbf{r}, t)$ at each point $\mathbf{r}$ in a medium. These quantities behave differently under time reversal, and it is obvious that

$$\rho(\mathbf{r}, t) \to \rho(\mathbf{r}, -t), \qquad \mathbf{j}(\mathbf{r}, t) \to -\mathbf{j}(\mathbf{r}, -t). \tag{3.76}$$

---

† This result is not altered even if the particles are identical and obey Fermi–Dirac or Bose–Einstein statistics, so that the state of the system is described by a suitably symmetrized sum of the products of one-particle wave functions [as shown in Eq. (88)], since the sum of a number of terms all having the same parity will also possess this parity. Furthermore, the spin of electrons, for instance, will also not affect this result directly, since the parity of the spin function can be chosen to be even.

The invariance of a system under time reversal has the following physical significance. Let a system be in the state A_0 at some initial instant of time t_0, and in the state A_1 at the subsequent time $t_0 + t_1$. Then if one chooses the instant of time $t_0 + t_1$ and the state A_1 as the initial ones, after a time t_1 the system must return to the state A_0. For instance, let us consider the motion of a positively charged particle in a magnetic field, as shown in Fig. 5. We choose the direction of the magnetic field **B** to be perpendicular to the plane of the paper and point into the paper, so that the charge rotates in the anticlockwise direction. If A_0 is the initial position of the particle, then after some time t_1 it will reach the point A_1. Let us now perform the time-reversal operation, i.e., choose the position A_1 as the initial one. When t is replaced by $-t$, the direction of the particle's velocity, which is along the tangent to its trajectory, will be reversed. A particle leaving A_1 with this reversed velocity would apparently arrive after time t_1 at the point A_2, rather than at its original starting point A_0. However, this conclusion is false, because we have implicitly assumed that time reversal does not affect the direction of the magnetic field. In order to remove the contradiction with that expected time-reversal symmetry of the system, we must assume that the magnetic field is reversed by time reversal, in which case the particle will return to A_0.

In order to convince ourselves that this is in fact the case, let us consider the motion of a particle of mass m and charge q, instantaneously at point **r**, moving under the influence of the Lorentz force, in an electric field $\mathbf{E}(\mathbf{r}, t)$ and a magnetic field $\mathbf{B}(\mathbf{r}, t)$. The particle's equation of motion (in Gaussian units) is

$$m\ddot{\mathbf{r}} = q\mathbf{E}(\mathbf{r}, t) + q\mathbf{v} \times \mathbf{B}(\mathbf{r}, t)/c. \qquad (3.77)$$

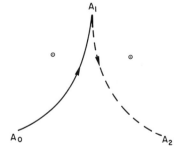

Fig. 5. Actual $(A_0 A_1 A_0)$ and hypothetical $(A_0 A_1 A_2)$ trajectories of a positively charged particle in a magnetic field pointing into the paper initially, with time inversion when the particle reaches A_1.

Since, under time inversion, $\mathbf{r}$ is unaltered while $\mathbf{v} \to -\mathbf{v}$, this equation will be invariant under time inversion if and only if

$$\mathbf{E}(\mathbf{r}, t) \to \mathbf{E}(\mathbf{r}, -t), \qquad \mathbf{B}(\mathbf{r}, t) \to -\mathbf{B}(\mathbf{r}, -t). \tag{3.78}$$

The fundamental reason for Eq. (78) is as follows. According to Maxwell's equations, the electric field is produced by a distribution of static charges $\rho(\mathbf{r}, t)$ that is unaffected by time inversion in accordance with Eq. (76). The magnetic field, on the other hand, is created by a distribution of moving electrical charges $\mathbf{j}(\mathbf{r}, t)$, whose sign is reversed by time inversion according to Eq. (76), and not by static magnetic charges, for instance.

The requirement of symmetry under spatial inversion, which led to the conservation of parity, enables us to derive from Eq. (77) the parity of the electric and magnetic field. Since $\mathbf{r} \to -\mathbf{r}$, and $\mathbf{v} \to -\mathbf{v}$ under spatial inversion, the equation will be invariant if and only if

$$\mathbf{E}(\mathbf{r}, t) \to -\mathbf{E}(-\mathbf{r}, t), \qquad \mathbf{B}(\mathbf{r}, t) \to \mathbf{B}(-\mathbf{r}, t). \tag{3.79}$$

Thus, the electric field has odd parity, and the magnetic field even parity. Finally, we note that under inversion of the coordinates of four-dimensional space–time in special relativity, i.e., both time reversal and spatial inversion, the electric and magnetic fields both change sign,

$$\mathbf{E}(\mathbf{r}, t) \to -\mathbf{E}(-\mathbf{r}, -t), \qquad \mathbf{B}(\mathbf{r}, t) \to -\mathbf{B}(-\mathbf{r}, -t). \tag{3.80}$$

This result is to be expected, as the components of these fields appear together as the elements of a second rank tensor in the relativistic formulation of Maxwell's equations.

Time-Reversal Symmetry and Irreversibility

An obvious discrepancy exists between the time-reversal symmetry of the basic equations of mechanics and the known irreversibility of most physical phenomena. For instance, as time passes, we all grow older and nobody has even seen the reverse process. The explanation of this irreversibility is that, in addition to the deterministic laws of mechanics that control the behavior of each individual particle, there are also probabilistic laws. These express the fact that every system tends to approach its most probable state, a law known as the principle of increasing entropy. A simple example is that of gas molecules enclosed in a volume V_1 that is connected by a valve with the larger volume V_2. When the valve is opened, each molecule can be in either volume, and the relative probability of its being in V_1 or V_2 is determined by the ratio of V_1 to V_2. Although trajectories exist that lead back from V_2 to V_1, their number is smaller by a factor V_1/V_2 than the number of trajectories that keep the molecule in the volume V_2 once it arrives there. As a result, after a short while, the probability of a given molecule being in volume V_1 is just $V_1/(V_1 + V_2)$, or $\frac{1}{3}$ if we choose V_2 to be twice the size of V_1. If V_1 initially contained 2 particles, the probability of their both returning to V_1 is $(\frac{1}{3})^2 = \frac{1}{9}$. However, if the volume contains some 10^{23} molecules, as is usually the case, the probability of their all returning simultaneously to V_1 is $(\frac{1}{3})^{10^{23}}$, i.e., it is negligibly small. In other words, the time-reversed situation allowed by the mechanical equations has a negligible probability of being realized in practice.

However, it can be shown that for some systems (called ergodic ones), after a sufficiently long time the system must pass through a state arbitrarily close to its initial one. An everyday example of this type of system is provided by a driver who has only a narrow garage for his car. Every driver knows that it is much easier to drive out of such a garage than to drive into it, as the proportion of successful car trajectories in the former case is much larger than in the latter. However, drivers are "ergodic" in their wish to safeguard their property, and so they eventually realize the event which has a smaller probability.

Time-Reversal Symmetry in Quantum Mechanics

The basic equation of quantum mechanics, Schrödinger's equation,

$$i\hbar\, \partial\psi(\mathbf{r}, t)/\partial t = \hat{H}\psi(\mathbf{r}, t), \tag{3.81}$$

involves a first derivative with respect to the time t, in contrast to the basic equations of classical mechanics. As a result, if we replace t by $-t$ in Eq. (81) and assume that $\hat{H}$ does not depend explicitily on t, we find that $\psi(\mathbf{r}, -t)$ satisfies

$$-i\hbar\, \partial\psi(\mathbf{r}, -t)/\partial t = \hat{H}\psi(\mathbf{r}, -t) \tag{3.82}$$

rather than Eq. (81). Thus, if time-reversal symmetry is to be maintained in quantum mechanics, the time-reversal operator $\hat{T}$ must have some effect on ψ in addition to replacing t by $-t$. A comparison of Eqs. (81) and (82) shows that $\psi^*(\mathbf{r}, -t)$ is a solution of Eq. (81), provided that the Hamiltonian operator $\hat{H}$ is real. This operation of complex conjugation does not affect the position of the particles described by the wave function ψ, since the particle density at any point is proportional to $\psi\psi^*$. Hence, a physically reasonable definition of the effect of $\hat{T}$ on a wave function, and one which leaves the Schrödinger equation invariant under time inversion,† is

$$\hat{T}\psi(\mathbf{r}, t) = \psi^*(\mathbf{r}, -t). \tag{3.83}$$

---

† The operator $\hat{T}$ defined by Eq. (83) is rather unusual, since it is neither linear nor unitary.

As we now show, this definition is in accord with our usual classical ideas. According to Eq. (83), if $\hat{A}$ is an arbitrary operator that does not depend explicitly on time, then $\hat{T}$ converts it into

$$(\hat{T}\hat{A})\hat{T}^{-1} = (\hat{A}^*\hat{T})T^{-1} = \hat{A}^*, \qquad \text{i.e.,} \quad \hat{T}\hat{A}\hat{T}^{-1} = \hat{A}^*. \qquad (3.84)$$

Two operators of this type are the operator of position $\hat{r}$ and that of momentum $\hat{p} = -i\hbar\nabla$. Hence, from Eq. (84),

$$\hat{T}\hat{r}\hat{T}^{-1} = \hat{r}, \qquad \hat{T}\hat{p}\hat{T}^{-1} = -\hat{p}, \qquad (3.85)$$

i.e., the time-reversal operator $\hat{T}$ leaves position unchanged but reverses the momentum, just as we would expect.

So far, we have considered only particles without spin. More general considerations, which are beyond the scope of this book, show that the behavior of a system of particles whose total spin is half-integral differs from that of a system whose total spin is an integer, because of their different behaviors under time reversal. It turns out that for systems with half-integral spin, but not for those with integral spin, the states described by ψ and $\hat{T}\psi$ are mutually orthogonal. Thus, they describe a pair of distinct degenerate states, and this degeneracy, known as Kramer's degeneracy, cannot be removed by a perturbation such as an electric field. It can only be removed by introducing into the Hamiltonian a term that does not possess time-reversal symmetry, e.g., a magnetic field, so that $\hat{H}$ and $\hat{T}$ no longer commute.

We should add two more comments about time-reversal symmetry in quantum mechanics. First, it does not lead to a conservation law for some quantity, because the operator $\hat{T}$ that is associated with it is not unitary. Second, the process of measurement, which is an integral part of quantum-mechanical theory, introduces a distinction between the past and the future, and so removes the time-reversal symmetry.

Indistinguishable Particles

The concept of indistinguishable particles, just like that of parity which we considered above, involves a symmetry transformation that leads to a new conservation law in quantum mechanics but not in classical mechanics.

The state of a particle in classical mechanics is characterized by its position and momentum. Since the equation of motion is a second-order differential equation, if the position and momentum of a particle are known at some instant t_0, its position and momentum can be found at an infinitesimal time later $t_0 + \delta t$, and by a repetition of this process the particle's trajectory can be derived. Similarly, for a system of particles, the trajectory of each particle can in principle be found, and if the particles are initially identified by the numbers $1, 2, \ldots, n$, we can find the position and momentum of any specific particle at any subsequent time.

In quantum mechanics, however, the situation is very different. First, there is Heisenberg's uncertainty principle, according to which the position and momentum of a particle cannot be specified simultaneously. Hence, if the position of a particle at a given instant is known exactly, its momentum is not known, and so its position an infinitesimal amount of time later does not have a definite value. Therefore, the concept of a particle's trajectory has no meaning in quantum mechanics. Second, if the position of a system of particles of the same type is specified at some instant, and at some subsequent time it is found that one of the particles is located at a given point, we cannot say which of the particles it is. For this reason, particles of the same type, i.e., which do not differ with regard to any of their characteristic parameters (mass, charge, etc.) are said in quantum mechanics to be "indistinguishable."

In order to analyze the consequences of indistinguishability, we consider first a system of two particles. Let the state of the system be described by the wave function $\Psi(\xi_1, \xi_2)$ where ξ_1 and ξ_2 are the dynamic variables, say the space coordinates and component of spin in the z-direction, of the two particles. We define the permutation operator $\hat{\Pi}_{12}$ by the equation

$$\hat{\Pi}_{12}\Psi(\xi_1, \xi_2) = \Psi(\xi_2, \xi_1). \tag{3.86}$$

The operator $\hat{\Pi}_{12}$ is a unitary operator which does not change the state of the system. Hence, it commutes with the Hamiltonian operator and specifies the existence of some new symmetry with an associated conservation law. On applying $\hat{\Pi}_{12}$ to Eq. (86) we find that

$$\hat{\Pi}_{12}^2\Psi(\xi_1, \xi_2) = \Psi(\xi_1, \xi_2). \tag{3.87}$$

Hence, the operator $\hat{\Pi}_{12}$ has just two possible eigenvalues ± 1. In other words, the wave function of a system of two particles must be either symmetric (i.e., unchanged) or antisymmetric (i.e., changed sign) with respect to the interchange of the particles.

If the mutual interaction of the particles can be neglected, the wave function Ψ of the system of two particles can be constructed from one-particle wave functions $\psi_a(\xi)$ and $\psi_b(\xi)$ associated with states a and b. When account is taken of the symmetry requirements with respect to the interchange of particles, the possible wave functions are

$$\Psi_\pm(\xi_1, \xi_2) = (1/\sqrt{2})[\psi_a(\xi_1)\psi_b(\xi_2) \pm \psi_a(\xi_2)\psi_b(\xi_1)]. \tag{3.88}$$

Here, Ψ_+ describes the symmetric and Ψ_- the antisymmetric state, while the factor $1/\sqrt{2}$ normalizes Ψ if $\psi_a(\xi)$ and $\psi_b(\xi)$ are normalized and mutually orthogonal. We see from Eq. (88) that if states a and b coincide, the antisymmetric function Ψ_- vanishes. In other words, for particles described by antisymmetric wave functions, no more than one particle can be in any given state, a rule known as Pauli's exclusion principle. Such particles obey Fermi–Dirac statistics, and so are usually called fermions. Similarly, particles that are described by symmetric wave functions are known as bosons, since they obey Bose–Einstein statistics. The statistical properties of fermions and bosons were discussed in Section 2.3.

An analogous description can be used for a system of N particles. Two types of state exist. For the completely symmetric state, the interchange of any pair of particles i and j by the permutation operator $\hat{\Pi}_{ij}$ leaves the system's wave function unchanged, while for a completely antisymmetric state the effect of $\hat{\Pi}_{ij}$ is to multiply the wave function by -1. These two types of state are thus eigenfunctions of the operator $\hat{\Pi}_{ij}$ with eigenvalues ± 1. Since $\hat{\Pi}_{ij}$ commutes with the Hamiltonian operator $\hat{H}$, a system that is initially in a completely symmetric (or antisymmetric) state will always remain in that state. Thus, the invariance with respect to permutation of a many-particle system leads to a conservation of its permutation symmetry and hence of the statistics which the particles obey.

Why do only two states of a many-particle system, the completely symmetric and the completely antisymmetric, exist in nature? Existing theory cannot answer this question fully, and this is to be regarded rather as an experimental fact. However, relativistic quantum theory, together with certain assumptions, does provide a partial explanation. In particular, it emerges that particles possessing half-integer spin obey Fermi–Dirac statistics, while those with integer spin are governed by Bose–Einstein statistics. For complex particles consisting entirely of fermions or of bosons, the statistics which they obey is determined by the number and type of particles that form them. For instance, the ^{4}He atom contains six fermions having half-integer spin (2 neutrons, 2 protons, and 2 electrons), and so has integer spin and obeys Bose–Einstein statistics. On the other hand, the ^{3}He atom contains only five fermions (1 neutron, 2 protons, and 2 electrons), and so has half-integer spin and obeys Fermi–Dirac statistics. This difference in statistics is responsible for the great differences between the properties of liquid ^{4}He and liquid ^{3}He.

The fact that quantum mechanics does not permit the identification of the individual particles in a many-particle system was expressed by John Wheeler as follows: "I know why all electrons have the same charge and the same mass: because they are all *the same* electron."

Gauge Invariance and Charge Conservation

In classical electrodynamics, as we saw in Section 3.1, the law of conservation of electric charge is an immediate consequence of the so-called gauge invariance of the basic equations. In quantum field theory, conservation laws exist also for other charges, such as the baryon charge, lepton number, isotopic spin, and strangeness mentioned in Section 1.3. All of these charge conservation laws are the consequences of gauge invariance and the arbitrariness of the choice of phase of a wave function, as we now proceed to show for the case of electric charge.

The first stage in the derivation of charge conservation laws is to find a quantum-mechanical expression for the current density of particles, which we denote here by $\mathbf{j}$. The probability of finding a particle in a fixed volume Ω is just $\int_\Omega |\psi|^2 \, dv$, and the current density is related to this quantity by the particle conservation law

$$\int_\Omega \nabla \cdot \mathbf{j} \, dv = -\frac{\partial}{\partial t} \int_\Omega |\psi|^2 \, dv = -\int_\Omega \left(\psi^* \frac{\partial \psi}{\partial t} + \psi \frac{\partial \psi^*}{\partial t} \right) dv. \quad (3.89)$$

In the absence of external electromagnetic fields, the system's Hamiltonian operator $\hat{H}$ is of the form

$$\hat{H} = \hat{p}^2/2m + V(\mathbf{r}, t) = -(\hbar^2/2m)\nabla^2 + V(\mathbf{r}, t), \quad (3.90)$$

where $V(\mathbf{r}, t)$ is the potential energy associated with all the other fields and forces acting on the particle.† Hence, on transforming the integrand on the right-hand side of Eq. (89) by means of Schrödinger's equation, (81), and its complex conjugate, and making use of the fact that $\hat{H}$ is Hermitian, we readily find that the particle current density is given by

$$\mathbf{j} = (i\hbar/2m)(\psi \, \nabla\psi^* - \psi^* \, \nabla\psi). \quad (3.91)$$

The differential form of the conservation law (89) can be written as

$$\partial\rho/\partial t + \nabla \cdot \mathbf{j} = 0, \quad \text{where} \quad \rho = |\psi|^2. \quad (3.92)$$

Multiplication of Eq. (92) by the electric charge e of the particle leads to the law of charge conservation in quantum mechanics in the absence of an external electromagnetic field, since $e\rho$ and $e\mathbf{j}$ are, respectively, the electrical charge density and electrical current density.

We now consider the effect of an external electromagnetic field, described by the scalar and vector potential ϕ and $\mathbf{A}$ according to

$$\mathbf{B} = \nabla \times \mathbf{A}; \quad \mathbf{E} = -\nabla\phi - (1/c) \, \partial\mathbf{A}/\partial t. \quad (3.93)$$

The Hamiltonian operator $\hat{H}$ for a particle of electrical charge e in this field is, according to quantum mechanics,

$$\hat{H} = (1/2m)(-i\hbar \nabla - e\mathbf{A}/c)^2 + e\phi + V(\mathbf{r}, t). \quad (3.94)$$

---

† For simplicity, we consider only single-particle wave functions and Hamiltonians, but the results can readily be generalized.

Let us change the potentials by the so-called gauge transformation [cf. Eq. (15)]

$$\mathbf{A} \to \mathbf{A}' = \mathbf{A} + \nabla\chi, \qquad \phi \to \phi' = \phi - (1/c)\, \partial\chi/\partial t, \qquad (3.95)$$

where $\chi(\mathbf{r}, t)$ is an arbitrary function of the coordinates and of the time. In view of Eq. (93), this gauge transformation does not change the electric and magnetic fields, which are the physically observable quantities, in contrast to the potentials. However, the gauge transformation of Eq. (95) does change the Hamiltonian operator $\hat{H}$ of Eq. (94), and hence the particle's wave function ψ, which is a solution of Schrödinger's equation. A comparison of the operators $\hat{H}$ and $\hat{H}'$ appropriate to the two different sets of potentials shows that the corresponding wave functions are related by

$$\psi' = \exp[(ie/\hbar c)\chi(\mathbf{r}, t)]\psi. \qquad (3.96)$$

Thus, the effect of the gauge transformation (93) is only to change the phase of the wave function. However, one of the main features of quantum mechanics is that all observable quantities depend only on the square of the modulus of the wave function, $|\psi|^2$, and so do not depend on its phase. For instance, a particle's mean position and energy, which are determined by the matrix elements of $\mathbf{r}$ and $\hat{H}$, respectively, are observable, and are independent of the phase of ψ. On the other hand, while the matrix element of the momentum *is* changed by the gauge transformation, momentum is not an observable quantity (in contrast to velocity).

In the presence of an external electromagnetic field, $\hat{H}$ is given by Eq. (94) rather than by Eq. (90). When account is taken of the extra terms in $\hat{H}$ in the derivation of the particle current density from Eq. (89), we find that Eq. (91) is replaced by an equation for the electrical current density

$$e\mathbf{j}(\psi, \mathbf{A}) = (ie\hbar/2m)(\psi\,\nabla\psi^* - \psi^*\,\nabla\psi) - (e^2/mc)\mathbf{A}\psi\psi^*. \qquad (3.97)$$

The current density is a physical observable, and we readily find that the gauge transformation of Eq. (95), together with the replacement of ψ by ψ' in accordance with Eq. (96), leves $e\mathbf{j}$ unaltered. This result shows the connection between gauge invariance and the conservation law for the electric charge. A proof that charge conservation does, in fact, follow from gauge invariance can be obtained from relativistic quantum mechanics, just as we found it simplest in Section 3.1 to derive charge conservation in the framework of relativistic classical mechanics.

In an analogous manner, the gauge invariance of other fields can be regarded as leading to conservation laws for their appropriate charges. Gauge invariance then requires that physical observables be unchanged by a transformation of field operators $\phi(\mathbf{r})$ according to $\phi(\mathbf{r}) \to \exp(i\chi)\phi(\mathbf{r})$. For instance, the similarity of the neutron and the proton allows us to regard them as two different states of the same particle, which we call the nucleon. Gauge invariance leads to the conservation of some quantity, which we call the baryon charge for the neutron–proton field. In other cases, the gauge invariance of the appropriate field leads to the conservation of lepton charge, strangeness, isotopic spin, etc.

As we have seen, the invariance of the Schrödinger equation under gauge transformations and the charge conservation that follows from this invariance are connected with the impossibility of measuring the wave function ψ itself (as opposed to $|\psi|^2$). It seems probable that this situation is quite general. A symmetry implies that two or more states are equivalent, and so that it is not possible by any measurement to distinguish between them. Thus, it is quite likely that all conservation laws are associated with such symmetries and reflect the impossibility of measuring some value. For instance, the conservation of energy and momentum is associated with the impossibility of measuring an absolute position in time or space, which is a consequence of the homogeneity of time and space.

Charge Conjugation

As we mentioned in Section 1.3, a large number of elementary particles have been discovered in recent years. For each such particle, there exists an antiparticle that has the same mass but differs from the original particle in the sign of the appropriate charges, i.e., in the sign of the interaction with some external field. The simplest example of such a pair of particle and antiparticle is the electron and positron, which have opposite electrical charges. The existence of the positron was predicted theoretically in 1931 and proved experimentally in the following year. The properties of other such pairs can be explained in terms of the different sign of the electric charge, baryon number or charge, lepton number, strangeness, hypercharge, or isotopic spin. For instance, the sign of the charge of the antineutron differs from that of the neutron in the sense that an antineutron disintegrates into an antiproton (a particle with the mass of a proton and negative electrical charge) and a positron, while the neutron decomposes into a proton and an electron. There are some particles, such as the photon and the neutral π-meson, for which the particle and antiparticle are identical.

We now define the charge-conjugation operator $\hat{C}$ as a simultaneous transformation of all quantities (operators, wave functions, etc.) when each particle is replaced by its corresponding antiparticle. Since particles and antiparticles are distinguished only by the sign of their interaction with external fields, it is natural to assume that a closed neutral system should be invariant with respect to the operator $\hat{C}$. In such a case, by analogy with parity, there will exist some conserved quantity, the so-called charge parity, with eigenvalues ± 1.

Twenty-five years ago, nobody doubted that the inversion operation $\hat{P}$, charge conjugation $\hat{C}$, and the time-reversal operation $\hat{T}$ were independent symmetry operations that reflected the fundamental laws of nature. Of these, inversion symmetry and the consequent conservation of parity seemed the most obvious, since it means that nature shows no preference between right and left, or in other words that the behavior of a system and of its mirror reflection are governed by the same laws. But, as we mentioned in Section 3.1, experiments on β-decay show that parity is not conserved in this interaction. The results of these experiments can be explained in terms of a combined parity, by the assumption that a system is invariant only with respect to the product of transformations $\hat{C}$ and $\hat{P}$ but not for each of them separately. Thus a symmetry between left and right is preserved but only in conjunction with a change from particles to antiparticles. However, subsequent experiments (with K-mesons) showed that even this combined parity is not conserved. Instead, systems are found to be invariant only under the $\hat{C}\hat{P}\hat{T}$ transformation, i.e., the combination of changing particles into antiparticles ($\hat{C}$), spatial inversion ($\hat{P}$), and the time reversal operation ($\hat{T}$). For the strong and electromagnetic interactions, each one of $\hat{C}$, $\hat{P}$, and $\hat{T}$ is an independent symmetry operation, but for weak interactions only their product $\hat{C}\hat{P}\hat{T}$ is in general a symmetry operation.

At present, the correctness of $\hat{C}\hat{P}\hat{T}$-invariance has been proved for some special types of forces, which assume, for instance, that the interaction between the elementary particles is local (the Pauli–Luders theorem). The future theory of elementary particles has to explain what are the special features of weak interactions, and also what is the connection between the operations $\hat{P}$ and $\hat{T}$, which correspond to inversion in four-dimensional space–time, and the seemingly different operation $\hat{C}$ of charge conjugation.

3.3 Symmetry and the Microscopic Properties of Systems

Following our consideration in Section 3.2 of general symmetries, we now turn to the symmetry properties of specific systems and their consequences. We define the symmetry of a system by the set of operators that transform the coordinate axes (for coordinates of space, spin, and/or time) in such a way that the description of the system is unaltered. This set of operators forms a group G, which is called the symmetry group of the system, since if two operators each leave a system unchanged, their product will also not change it.† For instance, the symmetry group of an isolated atom contains all pure rotations of the coordinate axes, plus all such rotations combined with the operation of inversion. A group G containing a finite number h of operations is called a finite group of order h. Our analysis in this and the next section will be concerned mainly with such finite groups. The symmetry of the elementary particles, on the other hand, which we discussed in Section 1.3, is believed to be associated with continuous symmetry groups which contain an infinite number of elements; the symmetry group of the free atom is a continuous group of this type.

A great difference exists between the symmetry requirements for the macroscopic properties of a system and those for its microscopic properties. The macroscopic properties (at least for an isolated system in thermal equilibrium) must possess the full symmetry of the system, i.e., be invariant under all the operations of its symmetry group G. These macroscopic properties generally arise from a statistical average over some microscopic property, and the latter does not have to possess such symmetry. For instance, the electronic charge distribution of an isolated atom in its ground state must be spherically symmetric, since there is no preferred direction, but the individual electronic wave functions do not in general have such symmetry. Similarly, for an insulating crystal, while the thermal conductivity must have the full symmetry of the crystal, the phonons that are responsible for producing it are not invariant under all the operations of the crystal's symmetry group. And while the static dielectric constant of water is isotropic, i.e., has spherical symmetry, the individual water molecules most certainly do not possess this symmetry. In this section, we consider the effect of symmetry on the microscopic properties of systems, while in the next section, Section 3.4, we will discuss its effects on macroscopic properties.

The Symmetry of Eigenfunctions

The microscopic properties of a system can generally be described in terms of the eigenfunctions u_n of some operator, denoted by $\hat{A}$, which is invariant under the h operation P_j of this system's symmetry group G. Usually, the operator $\hat{A}$ will be the quantum-mechanical Hamiltonian $\hat{H}$, and the u_n the appropriate wave functions. Since $\hat{H}$ is invariant under the P_j for a general function f we can write

$$P_j(\hat{H}f) = \hat{H}(P_j f). \tag{3.98}$$

† A set of elements forms a group if the product of any pair of elements is defined, the product of any two elements of the set belongs to the set, and the set contains an identity element and the inverse of each of its elements. A subgroup is a subset of the elements of a group that is in itself a group.

Hence, if u_n is an eigenfunction with eigenvalue λ_n, so that

$$\hat{H}u_n = \lambda_n u_n, \tag{3.99}$$

then

$$\hat{H}(P_j u_n) = P_j(\hat{H}u_n) = \lambda_n P_j u_n, \tag{3.100}$$

so that $P_j u_n$ is an eigenfunction degenerate with u_n.

For the sake of simplicity, let us start by considering a nondegenerate eigenfunction u_n of $\hat{H}$. Since it is nondegenerate, $P_j u_n$ must be a multiple of u_n, so that we can write

$$P_j u_n = c_{jn} u_n, \tag{3.101}$$

where c_{jn} is some constant.† It follows that

$$P_k P_j u_n = c_{kn} c_{jn} u_n, \tag{3.102}$$

so that if $P_k P_j = E$, the identity operation, then $c_{kn} c_{jn} = 1$. For instance, if P_j denotes a rotation of $180°$, a reflection, or an inversion of the coordinates, $P_j P_j = E$ and so $c_{jn}^2 = 1$. Thus, in this case $c_{jn} = \pm 1$, and we can classify u_n as odd or even with respect to the operation P_j according to whether $c_{jn} = -1$ or $+1$, respectively. The symmetry of u_n is determined by the set of coefficients c_{jn} for all the operations P_j belonging to G. In the language of group theory, these are said to specify the representation of G to which the function u_n belongs. In particular, a function that is invariant under all the operations of G, so that $c_{jn} = 1$ for all j, is said to belong to the identical representation of G.

Matrix Elements and Selection Rules

One very important application of the symmetry properties of eigen-functions is in the derivation of selection rules. While these can be derived and used for classical systems, as for the CO_2 molecule considered in Section 3.1, they are usually considered with respect to the wave functions of a quantum-mechanical system, and this is the approach that we will adopt. We consider a system whose stationary states are described in terms of non-degenerate single-particle wave functions u_p. We denote by M_{pq} the matrix element of an operator $\hat{M}$ between two such functions, i.e.

$$M_{pq} = \langle u_p | \hat{M} | u_q \rangle = \int u_p^*(\mathbf{r}) \hat{M} u_q(\mathbf{r}) \, dv. \tag{3.103}$$

---

† In this paragraph there is no summation over n.

Since a symmetry transformation of the integrand does not affect the value of an integral, either $u_p \hat{M} u_q$ belongs to the identical representation or M_{pq} is zero. According to the "golden rule formula," the transition probability per unit time from state q to state p, as a result of a perturbation characterized by the operator $\hat{M}$, is proportional to $|M_{pq}|^2$. Thus, the symmetry of the wave functions u_p, u_q and of the operator $\hat{M}$ will determine whether or not M_{pq} is zero and so whether the transition from state q to state p is permitted or forbidden. A simple example of this, in connection with inversion symmetry and parity, was considered in Section 3.2.

As a less trivial example of the derivation of selection rules, we consider the transitions induced by the electric field of a polarized ray of light between the electronic states of a system having symmetry group D_2. This group contains four symmetry operations, namely the identity $P_1 = E$, and rotations of 180° about the x, y, and z axes, which we denote by P_2, P_3, and P_4, respectively. Since $P_j^2 = E$ for all j, while $P_2 P_3 = P_4$, the only possible sets of values of the c_{jn} are those shown in Table 2. With each set, we list a function $u_n(\mathbf{r})$ for which these c_{jn} are the coefficients defined by Eq. (101); in these, $r = |\mathbf{r}|$ denotes the distance of a point from the origin.

Table 2

Representations of the Group D_2

Function u_n	Operation P_j:	P_1	P_2	P_3	P_4
$u_1 = rf(r)$	$c_{j1} =$	1	1	1	1
$u_2 = xf(r)$	$c_{j2} =$	1	1	-1	-1
$u_3 = yf(r)$	$c_{j3} =$	1	-1	1	-1
$u_4 = zf(r)$	$c_{j4} =$	1	-1	-1	1

Suppose now that an electric field E is applied in the x-direction, so that $\hat{M}$ is proportional to $\hat{x}E$ if we consider only dipole moments.† Then the effects of the operations P_j on the product $u_p \hat{M} u_q$ can readily be calculated from Table 2. For instance, $P_2(u_2 \hat{M} u_3) = -u_2 \hat{M} u_3$, so that, on integrating, we find that $M_{23} = -M_{23}$. Hence, M_{23} must equal zero. This is an explicit example of our previous statement that, if $u_p M u_q$ does not belong to the identical representation, then $M_{pq} = 0$. An examination of all the possible combinations shows that only $u_1 \hat{M} u_2$ and $u_3 \hat{M} u_4$ belong to the identical representation. Hence, only M_{12} and M_{34} can be nonzero, and so the only transitions that can be induced by a light wave with its electric field in the x direction (in the electric dipole approximation) are those between states

---

† The symmetry of the operator $\hat{x}$ is the same as that of the function x, and so of u_2.

having the symmetries of u_1 and u_2 or of u_3 and u_4. Thus, symmetry considerations reduce from six to two the number of possible transitions between states of different symmetry and also forbid transitions between states of the same symmetry.

Matrix elements such as M_{pq} are involved not only in the calculation of transition rates but also, for instance, in calculations of the eigenvalues of a system by means of perturbation theory or variational methods. In these cases, the symmetry of a system is of great value in simplifying the calculations.

*Irreducible Representations of Groups

Our treatment so far considered only nondegenerate eigenfunctions. In general, one or more of the functions $P_j u_n$ may not be proportional to u_n. In such a case, the set of operations P_j will generate from u_n a set of functions, which define a vector space A_n. This space can be spanned by a finite set of orthogonal functions v_k, i.e., each $P_j u_n$ can be expressed in a unique way as a linear combination of the functions v_k. The product $P_j P_k$ of any two elements of G belongs to G, since G is a group, and so the set of functions v_k must be closed under the operations of G. In other words, for each P_j there exists a set of numbers $R(P_j)_{kl}$ such that

$$P_j v_l = \sum_k v_k R(P_j)_{kl}. \tag{3.104}$$

The matrices $R(P_j)$ are said to form a representation of the group G generated by the set of basis functions v_k. The order of the matrices, which is the same as the number of basis functions, is called the dimension of the representation. If no subspace of the vector space A_n is closed under the operations of G, the representation is said to be irreducible. This will always be the case if the degeneracy of the eigenfunction u_n arises solely from the symmetry of the system. If u_n is nondegenerate, of course, the corresponding representation is irreducible, and in the notation of Eqs. (101) and (104),

$$R(P_j)_{11} = c_{jn}. \tag{3.105}$$

A nonsingular linear transformation of the functions v_k

$$v'_k = \sum_m v_m Q_{mk} \tag{3.106}$$

will produce a different set of basis functions v'_k and of matrices $R'(P_j)$. However, the sets of matrices R and R' will be related by a similarity transformation $R'(P_j) = Q^{-1} R(P_j) Q$. Representations which are related to each other by such a transformation are said to be equivalent. Thus, the functions associated with the vector space A_n belong to an irreducible representation that is equivalent to a uniquely defined one. It is shown in standard text-books on group theory that a finite group has only a finite number of non-equivalent irreducible representations. The reader who is interested is referred to these books for details of these, and of the selection rules for functions belonging to them. While the principles involved are the same as in our example, the details are more complicated.

One-Dimensional Representations

The full, complicated treatment of irreducible representations of groups is only required for groups in which not all the operators commute, i.e., $P_j P_k \neq P_k P_j$ for some pairs of elements. In many cases, however, the sym-metry group of a system may contain only elements that commute with each other, as was the case for the group D_2 considered in our example. Such a group is called Abelian. In other cases, useful results can be obtained by considering Abelian subgroups of the system's symmetry group. For Abelian groups, it is found that all the irreducible representations are one dimensional. Thus, the symmetry of a function is represented by a series of numbers, as in Eq. (101), rather than by the matrices R of Eq. (104). This property of Abelian groups is associated with the well-known fact that commuting operators can be diagonalized simultaneously, since Eq. (101) can be regarded as an eigen-value equation for the operator P_j. We note that the following sets of opera-tions commute with each other:

(i)   two rotations about the same axis;
(ii) rotations through 180° about mutually perpendicular axes;
(iii) reflections in mutually perpendicular planes;
(iv) a rotation and reflection in the plane perpendicular to the axis of rotation;
(v)   inversion and any rotation or reflection;
(vi) any two translations.

The Translational Symmetry of Crystals

A crystal is distinguished from other states of matter by its translational symmetry. As we discussed in Section 3.1, for any crystal it is possible to choose three non-coplanar basis vectors, which we again denote by $\mathbf{a}_1$, $\mathbf{a}_2$, and $\mathbf{a}_3$, such that translations through the vectors

$$\mathbf{n} = n_1 \mathbf{a}_1 + n_2 \mathbf{a}_2 + n_3 \mathbf{a}_3 \tag{3.107}$$

connect equivalent sites in the crystal. Strictly speaking, such a definition is only accurate for an infinite system or for one without surfaces. Otherwise, any such translation will alter the distance of a point from the crystal's surfaces and so change its environment. However, at points far away from the surfaces, their influence should be negligible, and the sites will be equivalent for all effects and purposes. In view of this, it proves convenient to make them exactly equivalent mathematically by introducing cyclic boundary conditions, which eliminate the surfaces. For these, we replace the actual crystal by a model one, in which translations of $N_1\mathbf{a}_1, N_2\mathbf{a}_2$, and $N_3\mathbf{a}_3$, where the N_j are large integers, each bring us back to the starting point. For a one-dimensional system, this is equivalent to replacing a straight line (which can be regarded as a circle of infinite radius) by a circle of large but finite radius. The validity of this procedure, provided that one is interested in bulk properties rather than in surface ones, has been established. The resulting set of translations form a finite group, called the translational group of the crystal. Since any two translations commute, this group is Abelian and all its irreducible representations are one dimensional.

We start by considering a one-dimensional crystal with basis vector a. A translation of the origin of coordinates through a, which we denote by P_a, must just multiply an eigenfunction $\psi(x)$ by a constant† c, $P_a\psi = c\psi$. Since a translation of Na returns us to the starting point,

$$P_a^N\psi(x) = \psi(x), \qquad (3.108)$$

and so

$$c^N = 1. \qquad (3.109)$$

Thus, the possible distinct values of c are just

$$c = \exp(-ika), \quad \text{where} \quad k = 2\pi r/Na, \quad -N/2 \le r < N/2, \quad r \text{ integer.} \qquad (3.110)$$

Hence

$$\psi(x - na) = P_a^n\psi(x) = \exp(-ikna)\psi(x), \qquad (3.111)$$

which is just Bloch's theorem in one dimension. Thus, the translational symmetry of any eigenfunction $\psi(x)$ is completely specified by the value of k appropriate to it.

---

† This is strictly true only for nondegenerate eigenfunctions. We leave it as an exercise for the reader to show that if $\psi(x)$ is degenerate, it can be decomposed into a sum of functions, each of which has this property for the appropriate value of c.

For a three-dimensional crystal, the natural generalization of the above results is

$$\psi(\mathbf{r} + \mathbf{n}) = \exp(i\mathbf{k} \cdot \mathbf{n})\psi(\mathbf{r}). \qquad (3.112)$$

The condition that each of the translations $N_1\mathbf{a}_1$, $N_2\mathbf{a}_2$, and $N_3\mathbf{a}_3$ returns us to the starting point leads to the requirement that

$$\exp(i\mathbf{k} \cdot N_1\mathbf{a}_1) = \exp(i\mathbf{k} \cdot N_2\mathbf{a}_2) = \exp(i\mathbf{k} \cdot N_3\mathbf{a}_3) = 1. \qquad (3.113)$$

In order to satisfy this requirement, we define a reciprocal lattice whose basis vectors $\mathbf{b}_j$ are such that

$$\mathbf{b}_p \cdot \mathbf{a}_q = 2\pi\delta_{p,q}. \qquad (3.114)$$

Then condition (113) is satisfied if

$$\mathbf{k} = (r_1/N_1)\mathbf{b}_1 + (r_2/N_2)\mathbf{b}_2 + (r_3/N_3)\mathbf{b}_3. \qquad (3.115)$$

We note that $\exp(i\mathbf{b}_j \cdot \mathbf{n}) = 1$ for all $\mathbf{n}$. Hence, if we define $\mathbf{K}$ as a translational vector of the reciprocal lattice,

$$\mathbf{K} = K_1\mathbf{b}_1 + K_2\mathbf{b}_2 + K_3\mathbf{b}_3, \qquad K_1, K_2, \text{ and } K_3 \text{ integers}, \quad (3.116)$$

the vectors $\mathbf{k}$ and $\mathbf{k} + \mathbf{K}$ correspond to the same translational symmetry, and so are said to be equivalent. All the possible distinct values of $\mathbf{k}$ can be obtained by requiring $-N_j/2 \leq r_j < N_j/2$ in Eq. (115), or by choosing all the points satisfying (115) that are closer to the origin than to any lattice point $\mathbf{K}$ of the reciprocal lattice. This latter choice, which is known as the first Brillouin zone, always has the full point symmetry of the crystal lattice, and so is generally used to define the range of distinct wave vectors $\mathbf{k}$.

Selection Rules for Crystals

Our first application of the translational symmetry is to the derivation of selection rules. Let $\hat{M}^q(\mathbf{r})$ denote an operator such that

$$\hat{M}^q(\mathbf{r} + \mathbf{n}) = \exp(i\mathbf{q} \cdot \mathbf{n})\hat{M}^q(\mathbf{r}). \qquad (3.117)$$

In analogy to Eq. (103), we define its matrix element between two wave functions $\phi_\mathbf{k}$, $\psi_{\mathbf{k}'}$ (whose translational symmetry is characterized by $\mathbf{k}$ and $\mathbf{k}'$) to be

$$M^q_{\mathbf{k},\mathbf{k}'} = \langle \phi_\mathbf{k} | \hat{M}^q | \psi_{\mathbf{k}'} \rangle = \iiint \phi_\mathbf{k}^*(\mathbf{r})\hat{M}^q(\mathbf{r})\psi_{\mathbf{k}'}(\mathbf{r}) \, dv. \qquad (3.118)$$

The volume of integration is one cycle of the model crystal, i.e., a parallele-piped of sides $N_1\mathbf{a}_1$, $N_2\mathbf{a}_2$, and $N_3\mathbf{a}_3$. Since a translation by $\mathbf{n}$ cannot affect the value of the integral, this can only be nonzero if the integrand is unchanged by such a translation, i.e., if $\exp((i\mathbf{k}' + i\mathbf{q} - i\mathbf{k}) \cdot \mathbf{n}) = 1$. Thus, the only permitted transitions are those for which

$$\mathbf{k}' + \mathbf{q} - \mathbf{k} = 0 \quad \text{or} \quad \mathbf{K}, \tag{3.119}$$

where $\mathbf{K}$ is a reciprocal lattice vector defined by Eq. (116). In particular, if $\hat{M}^{\mathbf{q}}$ corresponds to the perturbation due to the electric field of a light wave of wavelength λ, then $\hat{M}^{\mathbf{q}}$ is proportional to $\mathbf{E}_0 \exp(i\mathbf{q} \cdot \mathbf{r} - i\omega t)$ where $|\mathbf{q}| = 2\pi/\lambda$. Since λ is very much larger than the length of a typical lattice vector $\mathbf{a}_j$ (typically, $\lambda \sim 6000$ Å, $a_j \sim 6$ Å), $\mathbf{q}$ is in this case negligible in Eq. (119), so that direct optical transitions are only possible between states having equivalent wave vectors $\mathbf{k}$. Optical transitions between states with non-equivalent wave vectors are only possible with the participation of phonons.

The quantity $\hbar\mathbf{k}$ is sometimes referred to as the crystal momentum and Eq. (119) as the conservation of crystal momentum. We see that this conservation law arises from the requirement of invariance with respect to translations through a lattice vector $\mathbf{n}$, just as the law of conservation of ordinary momentum discussed in Section 3.1 arises from the invariance of a system with respect to arbitrary translations. The vector $\mathbf{K}$ appears in Eq. (119) because the invariance in a crystal is only with respect to discrete translations.

*Irreducible Representations of a Crystal's Space Group

We now turn to the effects of rotational symmetry in addition to the crystal's translational symmetry. The classification of irreducible representations by a wave vector $\mathbf{k}$ is retained when the rotations P_j of the crystal's point group G_0 are introduced. For convenience, we restrict our attention to symmorphic space groups, i.e., those that do not contain glide planes or screw axes. For a general function $\psi_{\mathbf{k}}(\mathbf{r})$ of translational symmetry $\mathbf{k}$, it is convenient to write

$$\psi_{\mathbf{k}}(\mathbf{r}) = \exp(i\mathbf{k} \cdot \mathbf{r})u_{\mathbf{k}}(\mathbf{r}). \tag{3.120}$$

In order for Eq. (112) to be satisfied, we require that $u_{\mathbf{k}}(\mathbf{r} + \mathbf{n}) = u_{\mathbf{k}}(\mathbf{r})$, so that $u_{\mathbf{k}}(\mathbf{r})$ has the periodicity of the crystal lattice. A rotation P_j of the co-ordinate axes transforms the coordinates of a point $\mathbf{r}$ to $P_j^{-1}\mathbf{r}$, while $(\mathbf{k} \cdot P_j^{-1}\mathbf{r}) = (P_j\mathbf{k} \cdot \mathbf{r})$, since the scalar product is invariant under rotations. Hence,

$$P_j\psi_{\mathbf{k}}(\mathbf{r}) = \exp(iP_j\mathbf{k} \cdot \mathbf{r})u_{\mathbf{k}}(P_j^{-1}\mathbf{r}). \tag{3.121}$$

Thus, for a general wave vector $\mathbf{k}$, P_j generates from $\psi_{\mathbf{k}}$ a function which has a different wave vector $P_j\mathbf{k}$, and so is linearly independent of $\psi_{\mathbf{k}}$. The functions formed by the operators P_j thus form a basis for an irreducible representation of the crystal's space group of dimension h_0, where h_0 is the number of elements in the group G_0. Only if **k** has a certain amount of symmetry, so that $P_j \mathbf{k}$ and **k** are equivalent, will $\psi_\mathbf{k}(\mathbf{r})$ and $P_j \psi_\mathbf{k}(\mathbf{r})$ correspond to functions having equivalent **k**, and so not necessarily be linearly independent. The set of such P_j for a given **k** is called the small group of **k**, and denoted by $G_\mathbf{k}$, while the set of nonequivalent vectors $P_m \mathbf{k}$ is known as the star of **k**.

*Structural Phase Transitions in Crystals

The problem of continuous phase transitions in crystals, which we considered briefly at the end of Section 3.1, provides an excellent example of the value and power of symmetry considerations. For convenience, we restrict our attention initially to structural phase transitions and do not, for instance, consider here changes in magnetic ordering in ferrimagnetic or antiferromagnetic crystals. The transition between different crystal phases, such as that from body-centered cubic to face-centered cubic in iron, is usually associated with a sudden rearrangement of the crystal lattice, in which the state of the crystal changes discontinuously. The liquid–solid transition on freezing a liquid, which we considered in Section 3.1, is also of this type.

However, in crystals there is also a possibility of a continuous phase transition that involves a change of symmetry. For instance, at high temperatures barium titanate has a cubic lattice, but as the temperature is lowered below a certain critical temperature the relative positions of the atoms begin to change, and the unit cell acquires a tetragonal distortion. The amount of distortion, as represented by the relative difference in length of the sides of the unit cell, changes continuously. Any nonzero value of this distortion, however small, corresponds to a change of symmetry, so that this distortion can be used as the order parameter η of Section 3.1. Another type of continuous phase transition is provided by systems in which the length of the unit cell is doubled below a certain critical temperature, as a result of alternate atoms approaching and becoming further from each other. In this case, the difference in the distances AB and BC between three successive atoms A, B, C that belong to adjacent unit cells can be taken as the order parameter η. An important feature of all phase transitions of this type is that the symmetry group of the lower phase is a subgroup of that of the upper phase.

The analysis of these phase transitions now proceeds initially in a manner analogous to that of Section 3.1. The density of atoms (possibly of a prescribed type) at point **r** is expressed as

$$\rho(\mathbf{r}) = \rho_0(\mathbf{r}) + \delta\rho(\mathbf{r}), \tag{3.122}$$

where ρ_0 is that part of ρ which is invariant under (i.e., belongs to the identical representation of) the symmetry group G of the upper phase. The function $\delta\rho$ can be expressed as the sum of functions $f^{(n)}(\mathbf{r})$ belonging to the jth row of the nth irreducible representation of G. Thus we can write

$$\delta\rho(\mathbf{r}) = \sum_n' \sum_j c_j^{(n)} f_j^{(n)}(\mathbf{r}), \tag{3.123}$$

where the prime indicates that the sum over n excludes the identical representation. Because this is excluded, no linear function of $\delta\rho$ can belong to the identical representation of G. On the other hand, the free energy Φ, which is a function of ρ, must be invariant under G. Hence, Φ cannot involve any term that is linear in $\delta\rho$. In the neighborhood of the transition point we can expand Φ as a power series in $\delta\rho$ and hence in the coefficients $c_j^{(n)}$ of Eq. (123). According to group theory, for each irreducible representation only one second-order invariant exists, and this is a positive-definite form in the $c_j^{(n)}$ which can be reduced, by a suitable choice of the expansion functions $f_j^{(n)}$, to a sum of squares. Thus, we can write

$$\Phi = \Phi_0 + \eta^2 \sum_n' A^{(n)} \sum_j (c_j^{(n)})^2 + O(\eta^3). \tag{3.124}$$

In accordance with the discussion at the end of Section 3.1, the $A^{(n)}$, which are functions of the pressure p and temperature T, must all be positive above the transition point, while at least one of them must be zero at the transition point and become negative below it. Since two $A^{(n)}$ will not normally vanish simultaneously along a line of points in the p–T plane, we conclude that for a continuous phase transition to occur, only one of the $A^{(n)}$ will vanish at the transition point and become negative below it, while all the others will remain positive. As a result, for the minimum value of Φ, all the $c_j^{(n)}$ will vanish above the transition point, while below it only the $c_j^{(n)}$ associated with a specific irreducible representation will be nonzero. Thus, above the transition point $\rho(\mathbf{r})$ is invariant under G, as we postulated, while below it $\delta\rho$ belongs to a specific irreducible representation of G. The extra condition for a continuous transition, in accordance with Eq. (47) and the discussion that followed it, is that no third-order invariant can be constructed from the quantities $c_j^{(n)}$. This, in itself, provides a severe restriction on the possible forms of $\delta\rho$ in the lower phase, and so on the symmetry of this phase.

The discussion so far has been quite general, and made no use of the translational symmetry that characterizes crystals. As we have seen the irreducible representations of a crystal are specified by the wave vector $\mathbf{k}$ that describes their translational symmetry, together with the representation m of the small group $G_{\mathbf{k}}$. Thus, in place of $A^{(n)}$ in Eq. (124), we write $A^{(n)}(\mathbf{k})$, where $\mathbf{k}$ is a continuous variable, and the sum over j in this equation is over wave vectors in the star of $\mathbf{k}$ plus the rows of the representation m. However, since $\delta\rho$ must be real, for each term with wave vector $\mathbf{k}$ there must be a complex conjugate term with vector $-\mathbf{k}$. Thus, if the star of $\mathbf{k}$ does not contain $-\mathbf{k}$, we must extend the sum to include the star of $-\mathbf{k}$.† The physical significance of $\mathbf{k}$ is that it describes the translational symmetry of the distortion that leads to the lower phase. For instance, a distortion that is the same in each unit cell is associated with $\mathbf{k} = 0$, while one that leads to a doubling of the unit cell in the direction $\mathbf{a}_1$ is characterized by $\mathbf{k} = \frac{1}{2}\mathbf{b}_1$.

A phase transition can only occur at a given wave vector $\mathbf{k}_0$ if $A^{(m)}(\mathbf{k})$ does not vanish first at any other wave vector $\mathbf{k}$. Since $\mathbf{k}$ is a continuous variable, this means that $A^{(m)}(\mathbf{k})$ must have a minimum at $\mathbf{k}_0$, a requirement that imposes severe restrictions on the possible values of $\mathbf{k}_0$. A convenient way of examining these is as follows. A function $\delta\rho(\mathbf{r})$ with $\mathbf{k} = \mathbf{k}_0 + \boldsymbol{\kappa}$, for small $\boldsymbol{\kappa}$, can be represented in terms of functions f_j having translational symmetry $\mathbf{k}_0$ plus coefficients c_j which have translational symmetry $\boldsymbol{\kappa}$, and so are slowly varying functions of the position coordinates $\mathbf{r} = (x_1, x_2, x_3)$. The free energy Φ will then be a function of the coefficients c_j and their spatial derivatives, and the crystal's most stable state will be that for which the integral of Φ over a unit volume is a minimum. Thus, a state with wave vector $\mathbf{k}_0$ can only be stable if this integral does not involve terms of first order in $\partial c_j / \partial x_p$. The only possible invariant combinations involving these terms are of the form

$$I_n = \sum_{j,k,l} a_{jkl}^{(n)} \int \left[c_j \frac{\partial c_k}{\partial x_l} - c_k \frac{\partial c_j}{\partial x_l} \right] dv. \qquad (3.125)$$

If no such I_n exists, a phase transition is certainly possible for this value of $\mathbf{k}_0$. On the other hand, if one or more of them does exist, while they may vanish at a given $\mathbf{k}_0$ for some specific values of p and T, they cannot be expected to do so at the same $\mathbf{k}_0$ for any range of values of these parameters. Thus, a phase transition along a line of points in phase space would seem to be impossible if any such invariants I_n exist. However, such phase transitions have been observed, for instance, in magnetic systems, with the ordered phase containing a spiral arrangement of magnetic moments.

The explanation of this paradox is that phase transitions can occur in which the symmetry of the lower phase is the same for a range of values of p and T, even though the value of $\mathbf{k}_0$ varies with these parameters. For a spiral arrangement of magnetic moments, for instance, this change in $\mathbf{k}_0$ corresponds to a change in the pitch of the spiral in the ordered phase as

† In this case, although $A^{(m)}(\mathbf{k})$ and $A^{(m)}(-\mathbf{k})$ are associated with different irreducible representations of G, they can vanish simultaneously along a line of points in the p–T plane. The reason for this is that states having wave vectors $\mathbf{k}$ and $-\mathbf{k}$ are not entirely independent, but are connected by the time-reversal symmetry of the Schrödinger equation that we considered in Section 3.2.

p and T change. The point symmetry of the lower phase is that of $G'_{\mathbf{k}_0}$, the small group of $\mathbf{k}_0$, expanded, if necessary, to include also operations that convert $\mathbf{k}_0$ into $-\mathbf{k}_0$ or vectors equivalent to it. For a phase transition to be observable, this must be the same along a line of points in phase space. We define the number of degrees of freedom of a wave vector $\mathbf{k}$ according to how $\mathbf{k}$ can vary without altering $G'_{\mathbf{k}}$. A wave vector that can move anywhere in reciprocal space without altering $G'_{\mathbf{k}}$ is said to possess three degrees of freedom, one required to lie on a given symmetry plane has only two, one that can only move along a symmetry axis has just one, while a wave vector at a special point in the Brillouin zone, such that any change in $\mathbf{k}$ alters $G'_{\mathbf{k}}$, has no degrees of freedom. If there are m possible invariants I_n, the locus of points in reciprocal space at which they will all vanish will be three dimensional if $m = 0$, two dimensional if $m = 1$, a line if $m = 2$, and a point if $m = 3$, while if $m > 3$ they will not usually vanish simultaneously for any value of $\mathbf{k}$. Hence, if all the I_n vanish, for a given p and T, at some point $\mathbf{k}_0$, then as p and T are varied slightly the condition for them all to vanish at a point $\mathbf{k}$ having the same expanded small group as $\mathbf{k}_0$ is that the number of independent invariants I_n should not exceed the number of degrees of freedom of $\mathbf{k}_0$. This, then, is the correct criterion for the possibility of observing a phase transition at a given wave vector $\mathbf{k}_0$. The actual calculation of the number of possible invariants requires aspects of group theory that are beyond the scope of this book.

*Integrals over the First Brillouin Zone

Another interesting example of the application of crystal symmetry is provided by a method of calculating integrals in reciprocal space that has been proposed recently. A problem that arises quite often in solid state physics is to calculate the integral over the first Brillouin zone (i.e., over all distinct values of $\mathbf{k}$) of a function $g(\mathbf{k})$ that has the periodicity of the reciprocal lattice, so that

$$g(\mathbf{k} + \mathbf{K}) = g(\mathbf{k}). \tag{3.126}$$

Typical examples of the function $g(\mathbf{k})$ are the energy of an electron in a band, the electron density at different points in a crystal, and the total scattering rate out of a state as a result of the electron–phonon interaction. We denote the integral by

$$W = \int_{\mathrm{BZ}} g(\mathbf{k})\, d^3k \tag{3.127}$$

where the letters BZ denote that the integral is over the first Brillouin zone.

We note first that, since the first Brillouin zone has the full symmetry of the crystal point group G_0,

$$W = \int_{BZ} g(\mathbf{k}) \, d^3k = \int_{BZ} g(P_j\mathbf{k}) \, d^3k = \int_{BZ} \frac{1}{h_0} \sum_j g(P_j\mathbf{k}) \, d^3k, \quad (3.128)$$

where the P_j are the h_0 operations of G_0. Thus, W equals the integral over the first Brillouin zone of a function

$$f(\mathbf{k}) = \frac{1}{h_0} \sum_j g(P_j\mathbf{k}) \quad (3.129)$$

that belongs to the identical representation of G_0. From Eqs. (126) and (129), plus the fact that $P_j\mathbf{K}$ is a reciprocal lattice vector, we see that

$$f(\mathbf{k} + \mathbf{K}) = f(\mathbf{k}). \quad (3.130)$$

Hence, $f(\mathbf{k})$ can be expanded as a Fourier series

$$f(\mathbf{k}) = f_0 + \sum_{\mathbf{n} \neq 0} f_\mathbf{n} \exp(i\mathbf{k} \cdot \mathbf{n}). \quad (3.131)$$

Since $\int_{BZ} \exp(i\mathbf{k} \cdot \mathbf{n}) \, d^3k = 0$ if $\mathbf{n} \neq 0$, the integral W equals f_0 times the volume of the first Brillouin zone, so that the problem of calculating this integral reduces to that of evaluating f_0.

In order to take full advantage of the crystal's symmetry, we choose from all the nonzero lattice vectors $\mathbf{n}$ a subset $\mathbf{n}_t$ such that the sets of vectors $\{P_j\mathbf{n}_t\}$ cover all the lattice vectors $\mathbf{n}$, and no vector $\mathbf{n}$ belongs to two such sets. Since $P_j f(\mathbf{k}) = f(\mathbf{k})$, the coefficients $f_\mathbf{n}$ for all the $\mathbf{n}$ in a given set will be equal. Thus, we can write

$$f(\mathbf{k}) = f_0 + \sum_{t=1}^{Q} f_t A_t(\mathbf{k}) \quad (3.132)$$

where

$$A_t(\mathbf{k}) = \sum_j \exp(i\mathbf{k} \cdot P_j\mathbf{n}_t) = \sum_j \exp(iP_j^{-1}\mathbf{k} \cdot \mathbf{n}_t), \quad (3.133)$$

and the total number of distinct vectors $\mathbf{n}_t$ is Q; we arrange the labels t so that $0 < |\mathbf{n}_t| \leq |\mathbf{n}_{t+1}|$.

If a point $\mathbf{k}_0$ could be found such that $A_t(\mathbf{k}_0) = 0$ for all $t > 0$, then from Eq. (131) we would find that $f_0 = f(\mathbf{k}_0)$. However, no such point exists. Instead, we try to find sets of points $\mathbf{k}_m$ such that, for a given set of t,

$$\sum_m \alpha_m A_t(\mathbf{k}_m) = 0, \quad (3.134)$$

where the weighting coefficients α_m satisfy

$$\sum \alpha_m = 1. \quad (3.135)$$

If $f(\mathbf{k})$ is a smooth function of $\mathbf{k}$, we expect the coefficients f_t in Eq. (132) to decrease rapidly as t increases. We try to find a set of points $\mathbf{k}_m$, known as special points in the Brillouin zone, for which Eq. (134) is satisfied for $1 \leq t \leq N$. Then, from Eqs. (132) and (134), we can write

$$f_0 = \sum_m \alpha_m f(\mathbf{k}_m) - \sum_{t=N+1}^{Q} f_t \sum_m \alpha_m A_t(\mathbf{k}_m), \tag{3.136}$$

and for sufficiently large values of N the second term on the right-hand side of this equation should be negligible.

In order to construct a set of special points, we make use of the following property. Suppose that, for a given set of points $\mathbf{k}_m$,

$$\sum_m \alpha_m A_{t_1}(\mathbf{k}_m) = 0, \tag{3.137}$$

while for another $\mathbf{k}$ not in this set

$$A_{t_2}(\mathbf{k}) = 0. \tag{3.138}$$

Then, for $t = t_1$ and $t = t_2$,

$$\sum_m \alpha_m A_t(\mathbf{k}_m) A_t(\mathbf{k}) = 0. \tag{3.139}$$

Since

$$\sum_s \exp(iP_s \mathbf{k} \cdot \mathbf{n}_t) = \sum_s \exp(iP_s \mathbf{k} \cdot P_j \mathbf{n}_t), \tag{3.140}$$

on substituting from Eq. (133) into Eq. (139) we find that

$$\sum_j \sum_m \alpha_m \sum_s \exp(i(\mathbf{k}_m + P_s \mathbf{k}) \cdot P_j \mathbf{n}_t) = 0. \tag{3.141}$$

Thus, with a suitable choice of the weighting factors, the set of points $\mathbf{k}_m + P_j \mathbf{k}$ satisfies Eq. (134) for both $t = t_1$ and $t = t_2$. This shows us how to enlarge a set of special points so as to increase the range of t for which Eq. (134) is satisfied, and so to increase N in Eq. (136).

Our final application of symmetry to this problem lies in the choice of the first few lattice points. The points $\mathbf{k}_1$ and $\mathbf{k}_2$ should be chosen so as to minimize the number of distinct values $f(\mathbf{k}_1 + P_j \mathbf{k}_2)$ that have to be calculated. In this respect, it is important to note that $f(P_s \mathbf{k}) = f(\mathbf{k})$, since $f(\mathbf{k})$ belongs to the identical representation of G_0. For instance, for a simple cubic lattice, $\mathbf{k}_0 = \frac{1}{4}(\mathbf{b}_1 + \mathbf{b}_2 + \mathbf{b}_3)$ satisfies Eq. (138) for all lattice vectors $\mathbf{n}$ with at least one of the n_j odd. For even n_j that are not all multiples of 4, it is satisfied by $\mathbf{k} = \frac{1}{8}(\mathbf{b}_1 + \mathbf{b}_2 + \mathbf{b}_3)$. Although there are 48 different operations P_j in the cubic point group, we readily find that the set of vectors

$\mathbf{k}_0 + P_j\mathbf{k}$ involves only four distinct values of $f(\mathbf{k})$, and that the following set of points:

$$\mathbf{k}_1 = \tfrac{1}{8}(\mathbf{b}_1 + \mathbf{b}_2 + \mathbf{b}_3), \qquad \alpha_1 = \tfrac{1}{8};$$

$$\mathbf{k}_2 = \tfrac{3}{8}(\mathbf{b}_1 + \mathbf{b}_2 + \mathbf{b}_3), \qquad \alpha_2 = \tfrac{1}{8};$$

$$\mathbf{k}_3 = 3\mathbf{b}_1/8 + \mathbf{b}_2/8 + \mathbf{b}_3/8, \qquad \alpha_3 = \tfrac{3}{8};$$

$$\mathbf{k}_4 = 3\mathbf{b}_1/8 + 3\mathbf{b}_2/8 + \mathbf{b}_3/8, \qquad \alpha_4 = \tfrac{3}{8};$$

satisfies Eq. (134) except for lattice vectors for which all the n_j are multiples of 4. The first error occurs for $\mathbf{n} = 4\mathbf{a}_1$, which corresponds to the fourteenth shell of atoms away from the origin. The application of this method is a simple numerical exercise (see, for instance, Problem 3.3).

3.4 The Inversion Symmetry and Magnetic Symmetry of Crystal Properties

For the macroscopic properties of crystals, in contrast to the microscopic properties discussed in the last section, the translational symmetry is irrelevant. Instead, as we mentioned in Section 3.1, the symmetry of the macroscopic properties is determined by the crystal's point group, i.e., by its rotational symmetry, in accordance with Neumann's principle. A useful restatement of that principle is that no asymmetry can appear in the property tensor of a crystal unless that asymmetry is present in the crystal. In this section, we extend the considerations of Section 3.1 to include the symmetry operations of spatial inversion and time reversal which we discussed in Section 3.2. In particular, time-reversal symmetry is of crucial importance for magnetic crystals, since time reversal changes the sign of the magnetic field as we saw in Section 3.2. The great value of symmetry predictions can be seen from the fact that two phenomena which we discuss below, piezomagnetism and the magnetoelectric effect, were predicted theoretically and only afterwards found experimentally.

Inversion Symmetry—Polar and Axial Tensors

Our analysis of tensor symmetry in Section 3.1 was based on the assumption that the components of a tensor of rank h behave, under a rotation of the coordinates, like the products of the components of h vectors. While this is certainly true for proper rotations, it is not necessarily true for improper ones (i.e., ones that include inversion as well as a rotation) such as reflections.

To see this, let us consider the effect of spatial inversion, i.e., the transformation $\mathbf{r} \rightarrow -\mathbf{r}$, on the linear and angular momenta of a particle of mass m. The linear momentum $\mathbf{p} = m\,d\mathbf{r}/dt$ is changed into $-\mathbf{p}$ by such a transformation, but the angular momentum $\mathbf{L} = \mathbf{r} \times \mathbf{p}$ is unaltered. In order to distinguish between these two types of behavior, any set of components that behaves like $\mathbf{r}$ under proper rotations is called a vector, and a vector is said to be a polar if it changes sign under the inversion operation (i.e., if it has odd parity) and axial if it does not (i.e., if it has even parity). The above distinction can readily be extended to tensors. A three-dimensional Cartesian tensor of rank h has 3^h elements $T_{ij\ldots m}$ (where there are h subscripts, each of which can assume the value 1, 2, or 3). A rotation of the coordinate axes such that the position vector $\mathbf{r} = (x_1, x_2, x_3)$ is transformed into $\mathbf{r}'$ can be specified by the orthogonal matrix $A = (a_{jk})$ according to

$$x'_j = a_{jk} x_k. \tag{3.142}$$

Here, as is usual in the treatment of tensors, the summation convention is used. A tensor T of order h is said to be polar if, under such a rotation, it transforms into

$$T'_{ij\ldots m} = a_{ip} a_{jq} \cdots a_{mu} T_{pq\ldots u}, \tag{3.143}$$

while for an axial tensor T,

$$T'_{ij\ldots m} = \|A\| a_{ip} a_{jq} \cdots a_{mu} T_{pq\ldots u}. \tag{3.144}$$

Since for a proper rotation the determinant of A, $\|A\|$, equals $+1$, the behavior of polar and axial tensors only differs for improper rotations, where $\|A\| = -1$.

For axial tensors, the method of Section 3.1 can be used, but an extra factor of -1 has to be inserted when the effect of an improper rotation is considered. The distinction between axial and polar tensors is most obvious for centrosymmetric systems, i.e., those for which spatial inversion is a symmetry operation. Since such an inversion is an improper rotation that multiplies each coordinate by -1, its effect on a tensor of rank h is to multiply each element by $(-1)^h$ or $(-1)^{h+1}$, according as the tensor is polar or axial. However, according to Neumann's principle this symmetry operation must leave the elements of a tensor unchanged. Thus, in a centrosymmetric system, all polar tensors of odd rank and all axial tensors of even rank will be null, i.e., the effects described by them cannot occur. These are, of course, just the tensors that have odd parity. For other systems having improper rotations as symmetry elements, no such general conclusions can be drawn, and each case has to be considered individually. We now consider a simple, but typical, case.

Optical Activity

An example of an axial tensor of rank 2 is provided by the optical activity tensor g_{jk}. An optically active system is one in which the phase velocity of circularly polarized light for which the direction of the electric field advances to the right (i.e., clockwise) $v_\curvearrowright$, differs from that for such light in which the field advances to the left (anticlockwise) $v_\curvearrowleft$. The difference between these phase velocities, for light advancing in the direction (l_1, l_2, l_3), can be expressed as

$$v_\curvearrowright - v_\curvearrowleft = g_{jk} l_j l_k, \tag{3.145}$$

where g_{jk} is a symmetric second-rank tensor. Since inversion does not change the product $l_j l_k$, but does change clockwise into anticlockwise and so interchanges $v_\curvearrowleft$ and $v_\curvearrowright$, g_{jk} must be an axial tensor.

We now examine which elements of g_{jk} can differ from zero for two types of crystal that are not centrosymmetric and whose point groups contain just two elements, the identity E and a rotation A. First, we choose A to be a rotation of $180°$ about the z axis, so that $A(x, y, z) = (-x, -y, z)$. Since A is a proper rotation, the fact that g_{jk} is an axial tensor makes no difference, and g_{jk} behaves just as ε_{jk} was found to behave in Section 3.1. The operation A leaves x^2, y^2, z^2 and xy unaltered, but multiplies xz and yz by -1. Hence, $g_{13} = -g_{13}$ and $g_{23} = -g_{23}$, i.e., g_{13} and g_{23} vanish, but g_{11}, g_{22}, g_{33}, and g_{12} are arbitrary. For our second type of crystal, we choose A to be the operation of reflection in the xy plane so that $A(x, y, z) = (x, y, -z)$. This operation A has the same effect as previously on x^2, xy, xz, etc., but it is an improper rotation. Hence, for the axial tensor g_{jk}, the use of A together with Neumann's principle requires that $g_{11} = -g_{11} = 0$, $g_{22} = -g_{22} = 0$, $g_{33} = -g_{33} = 0$, $g_{12} = -g_{12} = 0$, and only g_{13} and g_{23} can differ from zero. Thus, while the permitted form of ε_{jk} is the same for these two types of crystal, that of g_{jk} is very different.

The above results can be summarized by dividing all the 32 crystal point groups into the following three sets, according to their properties with regard to spatial inversion:

(1)   the 11 pure point groups, which contain only proper rotations;

(2)   the 11 centrosymmetric point groups, for which inversion is a symmetry operation;

(3)   the 10 mixed point groups, which contain improper rotations (e.g., reflections) but not the inversion operation.

For crystals whose point groups are of the first type, there is no difference between polar and axial tensors, while for those with point groups of the second type, polar tensors of odd rank and axial tensors of even rank are null. For crystals with mixed point groups, a difference exists between polar and axial tensors, and each case has to be examined individually.

Time-Reversal Symmetry—i-Tensors and c-Tensors

Just as some physical properties are characterized, with regard to their behavior under spatial inversion, by polar tensors and others by axial tensors, so we must distinguish between the behavior of different types of tensors with respect to time reversal. Those tensors whose components are left invariant by the time-reversal operation $\hat{T}$ will be called i-tensors, while those for which $\hat{T}$ changes the signs of all the components will be termed c-tensors. For instance, $\hat{T}$ does not affect the electric field, and so the electric dipole moment of a system, but it does change the sign of the magnetic field, and so of a system's magnetic moment [cf. Eq. (78)]. Hence, pyroelectricity is characterized by a i-tensor, but the corresponding magnetic effect, pyromagnetism, is characterized by a c-tensor.

For a system in which $\hat{T}$ is a symmetry operation, Neumann's principle requires that $\hat{T}$ leaves unaltered the elements of any property tensor. In that case, all c-tensors will be null, just as tensors of odd order are null in a centrosymmetric system. As recently as 1953, it was believed that pyromagnetism had not been discovered because all systems do possess time-reversal symmetry, so that such an effect is forbidden. However, subsequent analysis and experiments have shown that this is not correct, as we discuss below.

Before considering this, we must point out the difference between the static equilibrium properties of a system and its transport or nonequilibrium properties with regard to time reversal. For instance, the current density **j** in a system is usually related to the electric field **E** according to the equation

$$j_p = \sigma_{pq} E_q, \tag{3.146}$$

where σ is the electrical conductivity of the system. Now the current density $\mathbf{j}(\mathbf{r}, t)$ is obviously antisymmetric under $\hat{T}$, while the electric field is time symmetric [cf. Eqs. (76) and (78)]. Hence the conductivity σ must be a c-tensor. However, it is not correct to conclude from Neumann's principle that the electrical conductivity of time-symmetric systems must vanish, so that electrical conduction is a forbidden effect. The reason is that transport effects involving finite conductivities are associated with an increase in entropy and the dissipation of energy, as discussed in Sections 3.2, 4.1, and 4.3, and so with a preferred direction of time. Similarly, a system not in equilibrium generally approaches equilibrium as time proceeds, and so has no time-reversal symmetry. Thus the use of time-reversal symmetry and the classification of property tensors as i-tensors or c-tensors is only meaningful in connection with static equilibrium properties.

Magnetic Systems

The key question that we must now consider is which types of system, if any, do not possess time-reversal symmetry. The existence in a system of a nonzero electric charge density $\rho(\mathbf{r}, t)$ will not affect its behavior with respect to the time-reversal operation $\hat{T}$, since $\hat{T}\rho(\mathbf{r}, t) = \rho(\mathbf{r}, -t) = \rho(\mathbf{r}, t)$, as we saw in Eq. (76). On the other hand, if a system contains a nonzero current density $\mathbf{j}(\mathbf{r}, t)$, this will affect its time-reversal symmetry, since $\hat{T}\mathbf{j}(\mathbf{r}, t) = -\mathbf{j}(\mathbf{r}, t)$. Of course, an isolated nonsuperconducting system in equilibrium cannot contain nonzero macroscopic currents, as the finite resistivity would lead to the dissipation of energy by such currents and so cause them to decay. However, on a microscopic level, e.g., round an atom or a molecule, a nonzero current density can exist, provided that $\int \mathbf{j}(\mathbf{r})\, dv$ over any macroscopic volume vanishes. These currents will give rise to a nonzero local density of magnetic moments $\mathbf{m}(\mathbf{r})$ proportional to $\int \mathbf{r} \times \mathbf{j}(\mathbf{r})\, dv$, and $\mathbf{m}$, like $\mathbf{j}$, will have its sign changed by time reversal. Similarly, the electron spins can produce a nonzero local magnetic moment, and we expect that the sign of this too will be changed by time reversal, since for all magnetic fields $\hat{T}H(\mathbf{r}, t) = -H(\mathbf{r}, t)$ in view of Eq. (78). Dirac's theory of electron spin in relativistic quantum mechanics shows that this is in fact the case.

In view of this, whether or not a system possesses time-reversal symmetry depends on the arrangement of any microscopic magnetic moments that may be present in it. In nonmagnetic (i.e., paramagnetic and diamagnetic) crystals and in the absence of an external magnetic field, these elementary moments are arranged randomly at any instant and change in a random manner with time, so that their average value vanishes. In magnetic materials, on the other hand, these microscopic moments are ordered and do not change rapidly or randomly with time. In an ideal ferromagnetic crystal, the elementary moments are mostly oriented parallel to each other. As a result, the value of the mean magnetic moment per unit cell is nonzero. In the simplest antiferromagnetic crystals, the elementary magnetic moments are also ordered, but with several different orientations such that the net magnetic moment in each unit cell is zero. Magnetic materials with more complicated structures also exist. For instance, in ferrimagnetic crystals, the elementary moments are ordered with two or more different orientations, but the net moment per unit cell is nonzero. Another complex structure which is found is the helical magnetic structure, in which the orientation of the elementary moments rotates as one proceeds along some crystal axis.

Magnetic Point Groups

The inclusion of the time-reversal operation $\hat{T}$ as a possible symmetry operation greatly increases the number of possible point and space groups for magnetic crystals. For such crystals, atoms of the same type and situated in similar positions, but with magnetic moments pointing in opposite directions, are no longer equivalent. If some rotation exchanges the positions of these atoms, then this rotation will not be a symmetry operation of the magnetic crystal unless it is combined with $\hat{T}$. For instance, for the set of four atoms shown in Fig. 6a a rotation of 180° about the axis EG is not a symmetry operation and neither is $\hat{T}$ (although they each would be were it not for the magnetic moments associated with the atoms A, B, C, and D, which are all of the same type). However, the combination of this rotation and $\hat{T}$ is a symmetry operation for the system. Instead of using arrows to show the directions of the magnetic moments, we can color black the areas in which these moments are in one direction and color white the places where they are in the opposite direction, as in Fig. 6b. In such a case, the operation $\hat{T}$ can be regarded as changing black into white and vice versa: such an approach makes it much easier to visualize the possible symmetry operations for more complicated systems.

The possibility that $\hat{T}$, in combination with some rotation, may be a symmetry operation increases both the number of possible point groups (from 32 to 122) and the number of possible space groups (from 230 to 1651) for crystals. The resulting groups are known as magnetic groups, colored groups (because of the use of black and white to depict the directions of the moments), or Shubnikow groups. The magnetic point groups can be divided into three types with regard to time-reversal symmetry, in a manner exactly analogous to the classification of point groups with respect to spatial inversion. These types are:

(1)   the 32 ordinary point groups, which contain only ordinary rotations, and no operation that involves $\hat{T}$;

(2)   the 32 gray point groups, for which $\hat{T}$ is itself a symmetry operation; these are called gray since for every operation that transforms white into

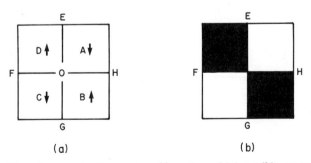

(a)   (b)

Fig. 6.   Magnetic symmetry for a system of four atoms. (a) A possible arrangement of the spins. (b) The corresponding black-and-white arrangement.

white, there is another symmetry operation ($\hat{T}$ times it) that converts white into black;

(3)   the 58 black and white point groups, for which $\hat{T}$ is not a symmetry operation but some combinations of $\hat{T}$ and rotations are.

Since $\hat{T}$ is a symmetry operation for nonmagnetic crystals, their point groups must be of type 2. In ferromagnetic and ferrimagnetic crystals, there is a preferred orientation of the magnetic moment, and so $\hat{T}$ cannot be an independent symmetry operation. Hence, the point groups of such crystals must be of type 1 or 3. For antiferromagnetic crystals, however, it turns out that the point group can be of any type.

When considering the tensor properties of crystals in connection with time reversal, we again make use of the analogy with spatial inversion. Just as all axial tensors of even rank are null for a crystal having a centrosymmetric point group, so all c-tensors are null for crystals that have a gray point group. And just as there is no difference between the behavior of polar and axial tensors for crystals whose point groups contain only proper rotations, so the distinction between i-tensors and c-tensors is irrelevant for magnetic crystals having an ordinary point group.

When account is taken of both inversion symmetry and time-reversal symmetry, we readily find the following set of rules for tensors associated with the static properties of crystals in equilibrium. Although some of these have been mentioned previously, we assemble them all here for convenience.

(a)   For nonmagnetic crystals, i.e., those having a gray point group, all c-tensors are null.

(b)   For a crystal whose point group contains either the spatial inversion operation $\hat{P}$ or this operation combined with the time-reversal operation $\hat{T}$, all i-tensors of odd parity, i.e., all polar i-tensors of odd rank and axial i-tensors of even rank, are null.

(c)   For magnetic crystals whose point group contains $\hat{P}$, polar c-tensors of odd rank and axial c-tensors of even rank are null.

(d)   For magnetic crystals whose point group contains $\hat{P}\hat{T}$, polar c-tensors of even rank and axial c-tensors of odd rank are null.

Pyromagnetism and Piezomagnetism

The phenomena of pyromagnetism and the piezomagnetic effect are analogous to those of pyroelectricity and the piezoelectric effect discussed in Section 3.1. Pyromagnetism consists of the appearance of a spontaneous, temperature-dependent, nonzero magnetic moment $\mathbf{M}°$ in a fixed direction in a crystal, while piezomagnetism involves the appearance of a magnetic moment as the result of the application of stresses. By analogy with Eq. (20), we can express the contribution of these effects to the free energy Φ in the form

$$\Phi_{\text{pyr}} = -(1/4\pi)B_j^0 H_j, \qquad \Phi_{\text{piez}} = -\alpha_{jkl} H_j \sigma_{kl} \qquad (3.147)$$

where $\mathbf{H}$ is the magnetic field in the crystal, and σ_{kl} the stress tensor. Since the free energy is an invariant, i.e., a polar i-tensor of rank zero, while $\mathbf{H}$ is an axial c-vector and σ_{jk} a polar i-tensor, B_j^0 and α_{jkl} are axial c-tensors of odd rank. Thus, neither of these effects can occur in crystals having gray point groups, nor in magnetic crystals whose point group contains the operation $\hat{P}\hat{T}$. In addition, pyromagnetism, just like pyroelectricity, requires the existence of a unique preferred axis, and so cannot occur, for instance, in cubic crystals. It is these symmetry restrictions, together with the small magnitude of the effects, that made them so difficult to observe. However, analysis of the relevant magnetic point groups shows that pyromagnetism is possible in the antiferromagnetic difluoride NiF_2, although not in the isomorphic compounds MnF_2, FeF_2, and CoF_2 because of the different orientations of the atomic spins in them, and this has subsequently been confirmed experimentally. Piezomagnetism is possible in all these crystals, and has been detected by a shift in the magnetization curve on the application of the appropriate stresses.

The Magnetoelectric Effect

The magnetoelectric effect consists of the appearance in a crystal of a magnetic induction $\mathbf{B}$ proportional to an external electric field $\mathbf{E}$, or of an electric displacement $\mathbf{D}$ proportional to an external magnetic field $\mathbf{H}$. The quantities $\mathbf{D}$ and $\mathbf{B}$ depend on the free energy according to its derivative with respect to the appropriate field, $\mathbf{D} = -(4\pi)\,\partial\Phi/\partial\mathbf{E}$ and $\mathbf{B} = -(4\pi)\,\partial\Phi/\partial\mathbf{H}$, and so this effect can occur only if the free energy contains a term of the form

$$\Phi_{me} = -q_{jk} H_j E_k. \qquad (3.148)$$

Since $\mathbf{H}$ is an invariant, while $\mathbf{H}$ and $\mathbf{E}$ are, respectively, an axial c-vector and a polar i-vector, q_{jk} is a second rank axial c-tensor. Hence, the magnetoelectric effect cannot occur in crystals whose point groups are gray or contain the inversion operation. It was first observed experimentally in 1960 in crystals of chromic oxide, Cr_2O_3, and we now proceed to derive the form of the tensor q_{jk} for such a crystal.

The unit cell of chromic oxide is rhombohedral, as shown in Fig. 7, with the chromium atoms situated along the trigonal axis. In the absence

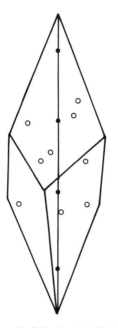

Fig. 7. The shape of the unit cell of Cr_2O_3; the filled circles denote Cr atoms, and the open ones O atoms. Note that the six O atoms on the side of the cell are each shared by two unit cells.

of magnetic moments, the point group of the crystal is D_{3d}, and contains the following 12 operations: E, the identity; $2C_3$, two rotations of 120°, clockwise and anticlockwise, about the long axis (which we choose as the z axis); $3C_2$, rotations of 180° about three axes perpendicular to the long axis; $\hat{P}$, the inversion; $2S_6$, two rotations of 60° about the long axis followed by reflections in a plane perpendicular to this axis; and $3\sigma_d$, reflections in three symmetry planes that contain the long axis.

In the antiferromagnetic phase of Cr_2O_3, the spins (and so the magnetic moments) of the four chromium atoms in a unit cell are in the $+z$ and $-z$ directions on alternate atoms. As a result, since spin is an axial vector, the improper rotations are only symmetry elements when combined with the time-reversal operation $\hat{T}$, so that the magnetic point group contains the elements E, $2C_3$, $3C_2$, $\hat{P}\hat{T}$, $2S_6\hat{T}$, and $3\sigma_d\hat{T}$. Incidentally, since this group contains $\hat{P}\hat{T}$, chromic oxide cannot exhibit piezomagnetism. It is immediately apparent that all the elements of the magnetic point group leave q_{33} unaltered, so that an electric field along the principal axis induces a magnetic moment in that direction. Since the symmetry group contains a threefold rotation, the analysis of the other elements of q_{jk} by the method of Section

3.1 is somewhat more complicated.† It is found that the only other nonzero elements of q_{jk} are q_{11} and q_{22}, and that these are equal, so that an electric field in the plane perpendicular to the principal axis will induce a magnetic moment parallel to itself. These predictions have all been confirmed experimentally, and the inverse effect, in which an applied magnetic field produces electric polarization, has also been observed.

The above examples show clearly the great power and importance of symmetry considerations in the prediction and analysis of physical phenomena.

Problems

3.1 A simple example of a non-Abelian group, i.e., one in which not all the operators commute, is provided by the symmetry group of the square, D_4. This consists of eight elements, namely, rotations of $0°, 90°, 180°$, and $270°$ about an axis perpendicular to the plane of the square and passing through its center, which we choose as the z axis, plus rotations of $180°$ about the lines that bisect opposite sides of the square (which we choose as the x and y axes) and about the square's diagonals.

(a) List the effect of all the operations of D_4 on the coordinates x, y, and z.

(b) Construct a table, similar to Table 2, for the effects of all these operations on the following functions, which can represent 3d orbitals:

$$u_1 = z^2 f(r), \qquad u_2 = (x^2 - y^2)f(r), \qquad u_3 = xyf(r),$$
$$u_4 = xzf(r), \qquad u_5 = yzf(r),$$

where r is the distance from the origin. Hence show that the functions u_1, u_2, and u_3 satisfy Eq. (101), and so generate one-dimensional representations of D_4, while u_4 and u_5 are the basis of a two-dimensional representation.

(c) Find which transitions between the states u_1, u_2, and u_3 can be induced, to first order in the field, by an electric field directed along the z axis.

---

† It follows from Eqs. (143) and (144) that an operation which changes x and y into linear combinations of each other, such as a rotation of $120°$, changes the elements of a tensor into the corresponding linear combination. For instance, an operation that converts x into $ax + by$ converts q_{11} into the combination of elements corresponding to $(ax + by)^2$, i.e., $a^2q_{11} + ab(q_{12} + q_{21}) + b^2q_{22}$, and for a symmetry operation this must equal q_{11}.

3.2 Prove that, because of the requirements of translational symmetry, the only proper rotations that can be symmetry operations for a crystal are those through an angle $2\pi/n$, where $n = 1, 2, 3, 4$, or 6. See Ref. [19, Section 132].

3.3 Calculate the mean value and root-mean-square value of the energy

$$E(\mathbf{k}) = E_0 + 2V(\cos k_1 a + \cos k_2 a + \cos k_3 a)$$

in the band of states $|k_j| \le \pi/a, j = 1, 2, 3$, for a simple cubic crystal
(a) by use of the special points listed at the end of Section 3.3, and
(b) by exact integration,
and compare the results. *Note:* The above form for $E(\mathbf{k})$ is that obtained in the tight-binding approximation for a simple cubic crystal.

3.4 (a) Show that the relationship $b_l = T_{lm}c_m$ between the components of the vectors $\mathbf{b}$ and $\mathbf{c}$ in a given Cartesian coordinate system will transform, under a rotation of the coordinate axes into $b'_l = T'_{lm}c'_m$, if and only if the elements T_{lm} are the components of a second rank tensor.
(b) Show that, if g is an axial scalar and $\mathbf{b}$, $\mathbf{c}$, and $\mathbf{d}$ are polar vectors, the relationship $g = A_{lmp}b_l c_m d_p$ is valid in all Cartesian coordinate systems if and only if A_{lmp} is an axial third-rank tensor.

3.5 The point group of graphite, which is denoted by 6mm, consists of rotations of $2r\pi/6$ about the z axis ($r = 0, 1, 2, 3, 4, 5$), plus six reflection planes containing the z axis, of which three pass through opposite vertices of a regular hexagon and the other three bisect opposite sides of it, as shown in Fig. 8.

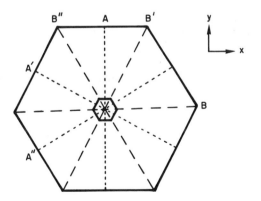

Fig. 8. The symmetry operations of the point group of graphite; A, A', A'', B, B', and B'' denote reflection planes containing the lines shown and the z axis.

(a) Show that all these operations can be obtained by repeated application of the following three operations, which are called the generators of the group:

A, reflection in the plane $x = 0$;

B, reflection in the plane $y = 0$;

C, a rotation of $2\pi/6$ about the z axis.

Derive the effects of these three operations on the coordinates x, y, and z.

(b) Show that for a crystal whose symmetry group is generated only by A and B, the off-diagonal elements of the dielectric tensor, ε_{jk} with $j \neq k$, are zero.

(c) Show that for graphite, the extra symmetry operation C leads to $\varepsilon_{11} = \varepsilon_{22}$.

(d) Which elements of the piezoelectric tensor γ_{jkl} vanish for graphite, and which nonzero elements are equal? Can graphite be pyroelectric?

3.6 The presence of a magnetic field **H** modifies the electrical resistivity tensor ρ_{jk}, and to second order in the field we can write

$$\rho_{jk}(H) = \rho_{jk}(0) + \alpha_{jkl} H_l + \beta_{jklm} H_l H_m,$$

where α_{jkl} and β_{jklm} are called the first-order Hall and magnetoresistance tensors, respectively. Note that, from Onsager's principle,

$$\rho_{jk}(-H) = \rho_{kj}(H).$$

Find, for each of these tensors, which elements are nonzero and which nonzero elements are equal for nonmagnetic crystals having the following point groups:

(a) S_2, which contains only the identity element and the inversion operator $\hat{P}$;

(b) D_4, the group described in Problem 3.1;

(c) C_{4v}, the group obtained from D_4 by replacing the rotations of $180°$ about axes in the xy plane by reflections in the planes containing these axes and the z axis;

(d) O, the simple cubic group, for which the symmetry operations produce all possible permutations of x, y, and z corresponding to proper rotations.

*3.7 As a comparatively simple example of the application of Onsager's principle, consider a system of two insulated containers of equal volume V, separated by a wall in which there is a hole of area A, and filled with an ideal monotonic gas. Let N denote the number of atoms, and U their total energy. We wish to calculate the fluxes of particles and energy, J_N and J_U, respectively, when small differences of pressure and temperature, Δp and ΔT, respectively, are introduced between the containers.

(a) For a system at constant volume, $T\,dS = dU - \mu\,dN$, where μ is the chemical potential per particle. Use this result plus the fact that for an ideal monatomic gas $\Delta(\mu/T) = V\Delta p/(NT) - 5R\,\Delta T/(2T)$ and $pV = NRT$ to derive the parameters X_1 and X_2 conjugate to $x_1 = N$ and $x_2 = U$. Hence show that, in terms of the parameters γ_{jk} of Eq. (26)

$$J_N = (\gamma_{11}R/p)\,\Delta p + (1/T^2)(\gamma_{12} - 5\gamma_{11}RT/2)\,\Delta T,$$

$$J_U = (\gamma_{21}R/p)\,\Delta p + (1/T^2)(\gamma_{22} - 5\gamma_{21}RT/2)\,\Delta T,$$

where p and T are the mean pressure and temperature.

(b) Assuming a Maxwell–Boltzmann velocity distribution for the gas atoms, calculate J_N and J_U directly, and confirm that Onsager's reciprocity principle, $\gamma_{21} = \gamma_{12}$, is satisfied. See (a) S. R. de Groot, "Thermodynamics of Irreversible Processes," Section 9. North-Holland Publ., Amsterdam, 1958, and (b) U. Lachish, *Amer. J. Phys.* **46**, 1163 (1978).

Chapter 4 | Analytical and Related Properties

4.1 Introduction

The analytical properties of physical quantities, just like their dimensionality and symmetry, enable us to study a number of general features of physical phenomena without the use of a detailed model and without inserting numerical values. The laws that describe physical phenomena prescribe how the magnitude of the quantity under consideration depends on the different factors that affect its behavior. Generally, the functions that describe this dependence are analytic functions of their arguments, and any breaking of analyticity must have a specific physical reason. In this chapter, we consider such analyticity requirements, and other general relationships that restrict the functional form of the dependence of quantities on parameters.

A function is analytic if it is differentiable to all orders at each point where it is defined, or equivalently if it can be expanded in a power series in the neighborhood of every such point. The points at which a function is not analytic are called its singular points or singularities. In physics, we are interested in several kinds of singular points, such as: poles, where a function tends to infinity; points of discontinuity; branch points, at which one can pass continuously from one set of values of the function (i.e., one branch) to another; and bifurcation points, at which a change occurs in the number of possible solutions of the equation defining the function. We start by considering some examples of the implications of analyticity and occurrence of singularities.

Restrictions on the values of parameters in a given problem, or on the applicability of a model, can sometimes be derived from the requirements of analyticity. For instance, as we saw in Section 2.1, the specific heat C of a material close to the critical temperature T_c varies with the temperature T as $(|T - T_c|/T_c)^{-\alpha}$, where α is a critical index. Experimentally, we measure an amount of heat $Q = \int C \, dT$ rather than the specific heat itself, and this must be finite. However, from our formulas, Q is proportional to $(T - T_c)^{1-\alpha}$, so that Q is finite only if α is less than unity. This result is quite general and can be used to test the validity of any model that leads to numerical values for the critical indices.†

As an example of how branch points can arise in physical problems, let us consider the normal modes of oscillation of atoms in a chain, which we discussed in Section 1.4. As we can see from Eq. (1.9), the frequency ω of the normal mode with wave vector q is related to q, at least for small q, by an equation of the form $\omega = cq$, where c is the velocity of propagation of the wave. However, since ω is a scalar whereas $\mathbf{q}$ is in general a vector, this formula must be written as $\omega = c\sqrt{q^2}$, so that ω, when regarded as a function of q^2, has a branch point at $q = 0$. The physical reason for the occurrence of this singularity is that the mechanical equations of motion for this system involve only second derivatives with respect to the time, and not first derivatives. Hence, their Fourier transform involves only ω^2, and not ω, so that ω^2 is required to be an analytic function of the scalar q^2, i.e., $\omega^2 = f(q^2)$. For small values of q, $\omega^2 = c^2 q^2$, and so $\omega = c\sqrt{q^2}$.

*Phase Transition Points

The best-known examples of bifurcation points are provided by phase transition points, which we have considered previously in Sections 2.1, 2.5, 3.1, and 3.3. The thermodynamic state of a system of N particles with pair interactions is determined by the partition function,

$$Z = \frac{1}{N!\Lambda^{3N}} \int d^3v_1 \cdots d^3v_n \exp\left(-\sum_{i<j} \frac{\phi(r_{ij})}{k_B T}\right), \qquad \Lambda = \sqrt{\frac{2\pi\hbar^2}{mk_B T}}. \quad (4.1)$$

Here, $\phi(r_{ij})$ is the interaction energy between particles i and j, distance r_{ij} apart, and m is the mass of a particle. Once we know Z, we can find the free energy F from the formula $F = -k_B T \ln Z$, the equation of state $p = p(V/N, T)$ from the thermodynamic relation $p = -(\partial F/\partial V)_T$, and so on.

The question that arises is how Eq. (1) can lead to a phase transition. The interaction energies ϕ do not, of course, depend on temperature. However,

---

† For the Ising model in the two-dimensional triangular lattice, calculations show that $\alpha = 0$ if only pair interactions between nearest neighbors are taken into account, but $\alpha = \frac{2}{3}$ if interactions between three adjacent spins are considered.

when the temperature is changed and approaches that of the phase transition, the particles "notice" this and start to arrange themselves on the sites of a crystal lattice for the transition from a liquid to a crystalline solid, or into two phases for the condensation of a vapor. The explanation of how this can follow from Eq. (1) is that a phase transition cannot occur in a finite system, but only in an infinite one in which N and V tend simultaneously to infinity with their ratio, the density $\rho = N/V$, remaining constant. In such a case, a bifurcation point can exist at which the number of solutions for N/V (i.e., ρ) increases. It was shown by van Hove in 1949 that the pressure p is a non-increasing function of the density ρ, so that Figs. 1a and b are possible forms for the graph of p against ρ. In Fig. 1b, when the external parameters T and p decrease (increase) to the point a (b), two values of ρ, ρ_a and ρ_b, satisfy the equation of state, i.e., two phases of different density coexist and so a phase transition occurs.

The van Hove theorem only indicates that Fig. 1b, and the associated transition, are mathematically possible. In order to obtain such a transition for a model system, it is necessary to choose a suitable interaction energy $\phi(r_{ij})$. In Chapter 1, we considered two models that led to phase transitions, namely the Ising model and the Kac model, which involve two limiting cases for the radius over which the interaction force extends.

A less well-known example of a bifurcation point arises in the problem of a three-component plasma of charged particles. Let us assume the existence of ionization equilibrium for the three components, which interact according to

$$AB \rightleftharpoons A^+ + B^-. \tag{4.2}$$

For instance, AB may be an alkali halide molecule, such as NaCl or KCl, which dissociates into ions A^+, B^-, as a result of the ionization reaction (2).

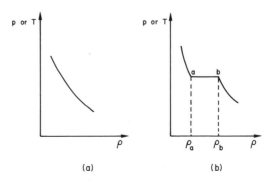

Fig. 1.   Possible forms for the dependence of the pressure p or temperature T of a system on its density ρ. (a) A system with no phase transition. (b) A system exhibiting a phase transition.

Another possibility is that AB represents a neutral atom, which dissociates into a positive ion A^+ and an electron B^-. Since the whole system is electrically neutral, the numbers of A and B charged particles must be equal, i.e., $N_A = N_B$. Moreover, the numbers of particles taking part in this chemical reaction (2) are connected by the law of mass action. In our case, this is of the form

$$\mu_{AB} = \mu_A + \mu_B, \tag{4.3}$$

where μ_j is the chemical potential of species j. For fluids, these chemical potentials are of the form

$$\mu_j = \mu_{j0}(p, T) + k_B T \ln(x_j \gamma_j), \tag{4.4}$$

where x_j is the number density of species j, and γ_j its activity which describes its deviation from an ideal system (for which $\gamma_j = 1$). Thus, the law of mass action (3) can be written

$$x_{AB}/(x_A x_B) = K_{id}(p, T)\gamma_A \gamma_B/\gamma_{AB} = K(p, T, x_j). \tag{4.5}$$

The chemical equilibrium constant $K_{id}(p, T)$ is determined by the functions $\mu_{j0}(p, T)$, i.e., by the properties of the individual components, while for a non-ideal system $K(p, T, x_j)$ also depends on the interactions between the components. The activities γ_j depend on the x_j, and for certain values of p and T, Eq. (5) can have more than one solution for the number of particles taking part in the reaction, i.e., a bifurcation point appears on the phase diagram. This means that this chemical reaction is "unstable." A first-order phase transition occurs (cf. Fig. 1b), and the system breaks up into two different phases, with different particle concentrations taking part in the chemical reaction (2).

Singularities and Analytical Relationships

While the branch points and bifurcation points that we have considered so far are of interest and importance, they are rather exceptional features of physical quantities. Other features of analyticity are the question of whether a quantity is finite or has a pole, and the relations between different quantities which place restrictions on their form. The rest of this chapter will be devoted to a systematic discussion of these features. According to Poincaré's theorem, the solution of a linear differential equation can have singularities in the finite plane only at points where the coefficients appearing in the equation are singular. Some of the consequences of this theorem will be considered in Section 4.2. Another important source of analytic properties and relationships is the requirement of causality, i.e., that the effect cannot precede the cause.

This leads to very useful dispersion relations, i.e., relationships between the frequency dependence of different physical parameters, as we discuss in Section 4.3. Finally, we consider in Section 4.4 the connection between the kinetic coefficients and correlation functions which can be derived from the fluctuation–dissipation theorem. We shall now discuss these topics in more detail.

Singularities in Quantum Mechanics

The quantum-mechanical wave function $\psi(\mathbf{r})$ is the solution of a linear differential equation, namely Schrödinger's equation. Hence, we can use Poincaré's theorem to study its analytical properties. For a particle moving in a spherically symmetric potential $V(r)$, for instance, and having angular quantum number l, Schrödinger's equation for the radial component $\psi(r)$ is

$$-\frac{\hbar^2}{2mr^2}\left[\frac{d}{dr}\left(r^2\frac{d\psi}{dr}\right) - l(l+1)\psi\right] + [V(r) - E]\psi = 0. \qquad (4.6)$$

The point $r = 0$ is the only singular point of the coefficients in this equation, and so the only point (in the finite plane normalization requirements exclude singularities as $r \to \infty$) at which ψ can have a singularity, provided that $V(r)$ is regular everywhere except at the origin. Whether it in fact does so depends on the form of the potential $V(r)$. If $V(r)$ is finite at the origin or tends to infinity less rapidly than r^{-2}, so that $\lim_{r\to 0} r^2V(r) = 0$, a solution of (6) exists with $\psi \sim r^l$ for small r. This solution is an analytic function of r at all points, including the origin. On the other hand, if $V(r)$ is negative near the origin and of form r^{-n} with $n > 2$, or with $n = 2$ and $r^2V(r)$ sufficiently negative, ψ will tend to infinity at the origin, with the nature of the singularity there depending on the detailed form of $V(r)$. This type of result is quite general. Singularities of the wave function in quantum mechanics can arise only from singularities of the potential. However, the converse is not true, and singularities of the potential need not lead to a singularity of the wave function there.

We now turn to the analytical properties of the scattering matrix, which was introduced by Heisenberg in 1943 to describe the asymptotic behavior of solutions of Eq. (6). It is convenient to define a wave number k in terms of the energy E of Eq. (6) by

$$E = \hbar^2k^2/2m. \qquad (4.7)$$

For a potential of finite range, the asymptotic solution of Eq. (6) for positive E has the form

$$\psi = (A/r)\sin(kr + \eta_l(k)), \qquad (4.8)$$

where the phase shift $\eta_1(k)$ is due to the potential $V(r)$. Equation (8) can be rewritten

$$\psi = (A'/r)[\exp(-ikr) - S(k)\exp(ikr)], \qquad S(k) = \exp(2i\eta_l(k)), \quad (4.9)$$

where A' is independent of r. The first term can be regarded as a wave incident on the origin, while the second one represents the scattered wave. Thus, the scattering matrix $S(k)$ describes the asymptotic behavior of the scattered part of a wave incident on the center. Since Eqs. (6) and (7) are invariant under the substitution $k \rightarrow -k$ and $\eta_l(k)$ is real, it must be an odd function of k. Hence, in view of (9),

$$S(-k) = S^*(k), \qquad S^*(k)S(k) = 1. \qquad (4.10)$$

These symmetry and unitarity properties of the S-matrix, which were derived here for a spherically symmetric potential, can be proved in general by use of the superposition principle and the law of conservation of the number of particles. Such properties are often useful for checking the validity of a model, as well as for simplifying calculations.

Some additional analytical properties of the S-matrix can be derived by regarding its argument, the wave number k, as a complex variable rather than a real one. Such a generalization of the domain of the argument enables us to use the very powerful mathematical tools of the theory of a complex variable, and in particular Cauchy's theorem of residues. According to this theorem, if $f(z)$ is analytical at all points within and on a simple smooth closed curve (or contour) C, apart from isolated singularities at the points $a_1, \ldots, a_M$ within C, then

$$\oint_c f(z)\,dz = 2\pi i \sum_{j=1}^{M} \sigma(a_j), \qquad (4.11)$$

where $\sigma(a_j)$ is the residue of $f(z)$ at the point a_j. It is found that the application of this procedure to $S(k)$ leads to a number of general analytical properties that follow from the causality principle, i.e., the requirement that the effect cannot precede the cause. In Section 4.2, we shall consider in some detail the analytic properties of the scattering matrix and the effects of causality on microscopic processes.

The Dielectric Constants of Model Systems

We now turn to the effect of causality on macroscopic phenomena. On proceeding, by means of a Fourier transform, from the relations of cause and effect in the time domain to the relationship between the corresponding quantities as functions of frequency (or energy), one finds that causality leads

to the so-called dispersion relations for physical properties. One very important feature of these relations is that they do not depend on the nature of the
interaction, but only on the fact that one phenomenon caused another. Hence,
they are especially useful in systems for which it is difficult to find a suitable
model for the interaction, such as the elementary particles which we discussed
in Section 1.3, and in systems where the model is too complicated to be solved,
such as the dielectric properties of ionic fluids.

Before we treat causality and dispersion relations, it is convenient to
define and derive the properties of a simple model system for which they are
of considerable importance. We therefore now consider briefly the dielectric
properties of a model solid in which the electrons are bound to their atoms by
damped harmonic forces with a natural frequency ω_0 and decay constant γ.
We assume, for convenience, that the local field at an atom produced by
surrounding atoms can be neglected, so that each electron–ion pair or
oscillator is affected only by the external electric field $E(t)$. If this is in the
x-direction, the equation of motion of such an oscillator is just

$$\ddot{x} + \gamma\dot{x} + \omega_0^2 x = (-e/m)E(t). \tag{4.12}$$

If we consider a periodic field $E(t) = E_0 \exp(-i\omega t)$ and write $x(t) = x_0 \exp(-i\omega t)$, the Fourier transform of Eq. (12) is

$$-\omega^2 x_0 - i\omega\gamma x_0 + \omega_0^2 x_0 = (-e/m)E_0, \tag{4.13}$$

so that

$$x_0 = (-e/m)E_0/(\omega_0^2 - \omega^2 - i\omega\gamma). \tag{4.14}$$

The polarization of the system $P(t)$ is determined by the displacement of the
electrons from their equilibrium positions, according to

$$P(t) = -enx(t), \tag{4.15}$$

where n is the electron density. Thus, on transferring to the frequency domain,

$$P_0 = -nex_0 = (ne^2/m)E_0/(\omega_0^2 - \omega^2 - i\omega\gamma). \tag{4.16}$$

The dielectric constant $\varepsilon(\omega)$ is defined by the relations

$$D_0 = E_0 + 4\pi P_0 = \varepsilon(\omega)E_0. \tag{4.17}$$

Hence, finally, we find that

$$\varepsilon(\omega) = 1 + (4\pi ne^2/m)/(\omega_0^2 - \omega^2 - i\gamma\omega). \tag{4.18}$$

The graph of the real part of $\varepsilon(\omega)$, as described by Eq. (18), is shown in
Fig. 2. The dotted line shows the value of $\varepsilon(\omega)$ if there is no decay, so that
$\gamma = 0$. In that case, $\varepsilon(\omega)$ has a point of discontinuity at $\omega = \omega_0$, approaching
$\pm\infty$ as $\omega \to \omega_0 \mp 0$. This resonance occurs at the natural frequency of the

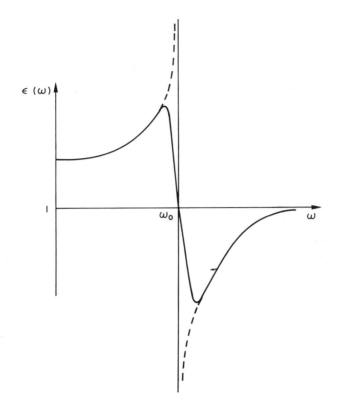

Fig. 2. The real part $\varepsilon'(\omega)$ of the dielectric constant $\varepsilon(\omega)$, as a function of the frequency ω, for a system of harmonic oscillators with no damping, and the full line to oscillators with damping.

oscillator, and this is the physical reason for the appearance of the singularity. However, if we take account of decay, so that γ differs from zero, we find that $\varepsilon(\omega)$ is an analytic function of ω along the real axis of frequency. We note that in the limit of high frequencies, i.e., $\omega \gg \omega_0$ and $\omega \gg \gamma$,

$$\varepsilon(\omega) = 1 - 4\pi n e^2/(m\omega^2). \tag{4.19}$$

For metals and other electrically conducting materials, it is convenient to extend the above definition of $\varepsilon(\omega)$, Eq. (17). To do this, we note that Maxwell's equation for $\nabla \times \mathbf{H}$,

$$\nabla \times \mathbf{H} = 4\pi \mathbf{j}/c + (1/c)\partial \mathbf{D}/\partial t \tag{4.20}$$

has, for insulators where $\mathbf{j} = 0$, the form

$$\nabla \times \mathbf{H}_0 = (-i\omega/c)\varepsilon(\omega)\mathbf{E}_0. \tag{4.21}$$

In order to preserve the relationship (21) for conductors, in which the dc current ($\omega = 0$) is given by

$$\mathbf{j} = \sigma \mathbf{E}, \tag{4.22}$$

we must in definition (17) make the replacement

$$\varepsilon(\omega) \rightarrow \varepsilon(\omega) + 4\pi i \sigma/\omega. \tag{4.23}$$

Dispersion Relations

We now turn to the general idea of causality and the dispersion relations to which it leads. Let an external time-dependent "input" $q(t)$ affect some physical system and lead to the appearance of an "output" $p(t)$ in the system. For instance, in the systems described above, an "input" consisting of the external electric field can be regarded as producing any of the "outputs" $x(t)$ of Eq. (12), $P(t)$ of Eq. (15), or $D(t) = E(t) + 4\pi P(t)$ of Eq. (17). We assume that the system is linear and additive, so that an input $q_1(t) + q_2(t)$ produces the output $p_1(t) + p_2(t)$, the sum of those that q_1 and q_2 would individually have produced. The system's response can then be characterized by a function $y(\tau)$ such that

$$p(t) = y_0 q(t) + \int_{-\infty}^{t} y(t - t')q(t')\, dt'$$

$$= y_0 q(t) + \int_{0}^{\infty} y(\tau)q(t - \tau)\, d\tau. \tag{4.24}$$

Here, we have picked out the first term for convenience, so that $y(\tau)$ need not include a δ-function $y_0 \delta(\tau)$. The second term takes account of the delay in producing a response to an input. The physical reason for this is that the molecular processes involved have finite relaxation times and cannot respond instantaneously. As a result, the value $p(t)$ of p at time t depends not only on the value $q(t)$ of q at that instant, but also on the values $q(t')$ of q at all previous instants t'. The causality principle states that, if q causes p, then $p(t)$ cannot depend on the value of q at any subsequent time $t' > t$, as the effect must precede the cause. Thus, $y(t - t') = 0$ for $t' > t$, or $y(\tau) = 0$ for $\tau < 0$, and this explains the choice of the limits of integration in Eq. (24).

On taking the Fourier transform of Eq. (24), using capital letters to denote the transforms of quantities represented by the corresponding lower-case letters, we find that

$$P(\omega) = Y(\omega)Q(\omega), \tag{4.25}$$

where

$$Y(\omega) = y_0 + \int_0^\infty y(\tau)e^{i\omega\tau}\,d\tau. \tag{4.26}$$

As in Eq. (24), so in (26) the lower limit of the integral is zero. The function $Y(\omega)$ is in general complex, and we write

$$Y(\omega) = Y'(\omega) + iY''(\omega). \tag{4.27}$$

Since $y(\tau)$ and y_0 must be real, we see from (26) that

$$Y(-\omega) = Y^*(\omega). \tag{4.28a}$$

so that Y' is an even function and Y'' an odd function of ω,

$$Y'(-\omega) = Y'(\omega), \qquad Y''(-\omega) = -Y''(\omega). \tag{4.28b}$$

We now proceed to derive a connection between the real and imaginary parts of $Y(\omega)$. Let us choose some real value ω_0 of ω, and integrate the function

$$f(\omega) = [Y(\omega) - y_0]/(\omega - \omega_0) \tag{4.29}$$

round the contour shown in Fig. 3. As can be seen from equation (26), $Y(\omega) - y_0$ is an analytic function of ω in the upper half plane $\omega'' > 0$ (where we write $\omega = \omega' + i\omega''$). Hence, by the residue theorem, we find that the principal value of the integral of $f(\omega)$ along the real axis, denoted by $\int_{-\infty}^\infty f(\omega)\,d\omega$, is just equal to minus the integral over the infinitesimal semi-circle round ω_0.† This is readily evaluated, and we find that

$$\int_{-\infty}^\infty \frac{Y(\omega) - y_0}{\omega - \omega_0}\,d\omega = i\pi[Y(\omega_0) - y_0]. \tag{4.30}$$

We now obtain dispersion relations by taking the real and imaginary parts of Eq. (30). On substituting x for ω and ω for ω_0, we find that

$$Y'(\omega) - y_0 = \frac{1}{\pi}\int_{-\infty}^\infty \frac{Y''(x)}{x - \omega}\,dx, \tag{4.31}$$

$$Y''(\omega) = -\frac{1}{\pi}\int_{-\infty}^\infty \frac{Y'(x) - y_0}{x - \omega}\,dx. \tag{4.32}$$

---

† We have assumed that the integral over the infinite semicircle vanishes, as it will if $Y(\omega) - y_0$ tends to zero like some negative power of ω when $|\omega| \to \infty$. This condition is normally fulfilled, because of the finite duration of the response to an input; cf. Eq. (19).

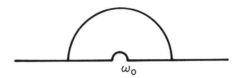

Fig. 3.   The contour used for integration to derive Eq. (30).

These general formulas, which are a direct result of the causality principle for a linear system, are called the Kramers–Kronig relations. We emphasize that they must be satisfied by the response function for any such system, whatever the microscopic mechanism that produces the response.

An additional term can appear in Eq. (32) if the function $Y(\omega)$ has a simple pole at $\omega = 0$, as does the dielectric susceptibility of a conductor defined in Eq. (23). If $Y(\omega) \sim iA/\omega$ near $\omega = 0$, an indentation to the contour consisting of an infinitesimal semicircle must be introduced at $\omega = 0$, similar to that at $\omega = \omega_0$, and the consequent contribution to the right-hand side of (30) will be $\pi A/\omega_0$. As a result, Eq. (32) becomes

$$Y''(\omega) = -\frac{1}{\pi} \int_{-\infty}^{\infty} \frac{Y'(x) - y_0}{x - \omega} \, dx + \frac{A}{\omega}. \qquad (4.33)$$

The above dispersion relations can be put into a more useful form if we split the integration over x into two parts, from $-\infty$ to 0 and from 0 to ∞, and use the parity relations of Eq. (28). We then find that

$$Y'(\omega) - y_0 = \frac{2}{\pi} \int_0^{\infty} \frac{xY''(x)}{x^2 - \omega^2} \, dx \qquad (4.34)$$

and

$$Y''(\omega) = \frac{A}{\omega} - \frac{2\omega}{\pi} \int_0^{\infty} \frac{Y'(x) - y_0}{x^2 - \omega^2} \, dx. \qquad (4.35)$$

As an example of the application of these formulas, let us consider the case where $Y(\omega)$ is the dielectric susceptibility $\varepsilon(\omega)$, defined by Eq. (17). We see at once that its limiting values (19) and (23), for $\omega \to \infty$ and $\omega \to 0$, respectively, satisfy the parity requirements of Eq. (28). For the harmonic oscillator model, in which $\varepsilon(\omega)$ is given by Eq. (18), the poles of $\varepsilon(\omega)$ are located in the lower half-plane of the complex variable ω, i.e., have $\omega'' < 0$, so that the analyticity requirements for integration over the contour of Fig. 3 are satisfied. Equation (19) ensures that $\varepsilon(\omega) - 1$ tends to zero rapidly enough as $|\omega| \to \infty$ for the integral over the infinite semicircle to vanish.

Hence, the Kramers–Kronig relations, Eq. (31) and (32) or (34) and (35), should apply to $\varepsilon(\omega)$, and the fact that this is so can readily be checked by direct integration.

Sum Rules

In practice, one of the main applications of dispersion relations is to the derivation of sum rules, which can readily be used in the analysis of experimental results. These will be considered in detail in Section 4.3, and we content ourselves here with one example. For this, we put $Y(\omega) = \varepsilon(\omega)$ and take the limit in Eq. (34) as $\omega \to \infty$. According to Eq. (19), in this limit $Y(\omega) - y_0 = \varepsilon(\omega) - 1 \sim -4\pi ne^2/m\omega^2$. Hence, we can neglect x in comparison to ω in the denominator of the integrand, and so find

$$\int_0^\infty x\varepsilon''(x)\, dx = \frac{2\pi^2 ne^2}{m}. \tag{4.36}$$

This equation does not involve any of the parameters of our model, viz., γ and ω_0, and so does not depend on any model [but only on Eq. (19)].

The imaginary part $\varepsilon''(\omega)$ of the dielectric constant is associated with the dissipation by the system of the energy supplied by the external field. Hence, it can differ from zero in general only for frequencies of the external field close to one of the characteristic frequencies of the system. Thus, to a good approximation (i.e., ignoring the finite widths of the absorption lines) we can write

$$\varepsilon'' = \frac{4\pi^2 ne^2}{m} \sum_k f_k \delta(\omega^2 - \omega_k^2), \tag{4.37}$$

where the ω_k are the characteristic frequencies and the f_k the corresponding "oscillator strengths." On substituting this form in Eq. (36), we obtain an equation

$$\sum_k f_k = 1, \tag{4.38}$$

which is known as a sum rule.

A somewhat different formulation of the sum rule can be derived for the harmonic oscillator model discussed above. If there are several types of electrons, n_k of which have characteristic frequency ω_k for $k = 1, 2, \ldots$, Eq. (18) will be replaced by

$$\varepsilon(\omega) = 1 + \frac{4\pi e^2}{m} \sum_k \frac{n_k}{(\omega_k^2 - \omega^2 - i\gamma_k\omega)}. \tag{4.39}$$

On taking the limit $\omega \to \infty$, we obtain a sum rule from Eq. (36) in the form

$$\sum_k n_k = n. \tag{4.40}$$

Since Eq. (40) is equivalent to Eq. (38) if we write $n_k = n f_k$, the meaning of the oscillator strength in the harmonic oscillator model is the fraction of electrons having characteristic frequency ω_k.

For quantum-mechanical systems, a sum rule analogous to (38) and (40) exists. However, instead of different sets of electrons with different natural frequencies, we have a single set of electrons which can be excited from their ground state, for instance, to various different excited states with corresponding different natural frequencies and oscillator strengths, associated, respectively, with the energy differences and transition probabilities. The corresponding sum rules will be discussed in Section 4.3.

The form of the causality principle discussed so far was a macroscopic one, that the cause must precede the effect, and from it we were able to derive equations such as (36) and (38) which can be directly compared with experiment. Other forms of this principle also exist. For instance, in the theory of special relativity it has the form "no signal can propagate with a velocity greater than that of light in vacuum." In quantum field theory, it is expressed by "local commutativity," i.e., the commutation of operators associated with spatially distinct points. These forms of the causality principle also have important consequences, but these are beyond the scope of our present treatment.

Causality and Time-Reversal Symmetry

Although it is not directly associated with analytic and related properties, it is worthwhile to consider the connection between the causality principle and the symmetry operation of time inversion, $t \to -t$. At first sight, it seems that the distinction between cause and effect implies an ordering of time, i.e., a preference for time to proceed in one direction rather than in the other, which contradicts the symmetry of the mechanical equations of motion under the time-reversal operation. However, such a conclusion is erroneous. If the order to time is reversed, $t \to -t$, what was previously an effect can now become a cause, and vice versa. In the model discussed above, Eq. (12) for the harmonic oscillator led to a function $\varepsilon(\omega)$, described by Eq. (18), that satisfies the causality principle and so the Kramers–Kronig relations only because of the friction term $\gamma \dot{x}$ in Eq. (12). However, friction lies, strictly speaking, outside the framework of mechanics, and the inclusion of this term means that Eq. (12) is no longer invariant under the transformation $t \to -t$. To remain within the limits of mechanical equations having time reversal symmetry, we must set $\gamma = 0$ in Eq. (12). The solution of the resulting equation $\ddot{x} + \omega_0^2 x = (-e/m)E(t)$ can be expressed in terms of $E(t)$ at either past times or at future times.† It is only friction that leads to an ordering of time. It achieves this by converting the ordered motion of the body into heat, i.e., a disordered motion of the particles of the surrounding medium, with a consequent increase in the entropy of the total system. A simple example of why such an increase in entropy leads to a preferred direction of time was presented in Section 3.2. This is an irreversible process of the type that we discussed in Sections 3.2 and 3.4.

*Fluctuations and Dissipation

The Kramers–Kronig relations, Eqs. (34) and (35), are examples of relationships between different physical properties which can affect the possible analytic form and even the analyticity of the functions describing them. Another important source of such relationships is the consideration of fluctuations in the properties of a many-particle system. These will be discussed in detail in Section 4.4, but as a simple example of what is involved we consider here the Brownian motion of a particle immersed in a fluid.

Under normal conditions, a particle immersed in a liquid will undergo some 10^{21} collisions per second with the molecules of the liquid. Fluctuations around this average number lead to a force responsible for the continuously changing motion of the immersed particle, which is called a Brownian particle. The resulting motion, like any forced motion, is opposed by a frictional force proportional to the velocity of the particle. As a result, the equation of motion of a Brownian particle of mass m can be written in the form

$$m\,du/dt = -m\gamma u + f(t), \tag{4.41}$$

where $u(t)$ is the particle's velocity and $m\gamma$ the friction constant. However, the two parts of the force acting on the Brownian particle, the systematic one $-m\gamma u$ and the random one $f(t)$, both arise from collisions with the molecules of the liquid, and so must be related. In fact, the fluctuation force $f(t)$ has the stochastic properties

$$\langle f(t)\rangle = 0, \qquad m\gamma = \frac{1}{k_{\rm B}T}\int_0^{\infty} \langle f(t_0)f(t_0+t)\rangle\,dt. \tag{4.42}$$

Thus, the systematic part of the force acting on a Brownian particle, namely the friction force, is determined by the correlation of the random force.

---

† Similarly, the electromagnetic fields produced by an accelerating charged particle can be described equally well either by a retarded potential, which depends on its past motion, or by an advanced potential, which depends on the particle's future motion.

The mean square value of the displacement of a Brownian particle in a given direction, $\langle[x(t) - x(0)]^2\rangle$, is a quantity that is observable experimentally and was first calculated by Einstein, in 1905. He found that

$$\lim_{t\to\infty} \langle[x(t) - x(0)]^2\rangle = 2Dt, \tag{4.43}$$

where D is called the coefficient of Brownian diffusion. This is related to the viscous friction of a Brownian particle by the Stokes–Einstein formula,

$$D = k_B T/m\gamma. \tag{4.44}$$

The displacement of a particle is determined by its velocity, since

$$x(t) - x(0) = \int_0^t u(t')\,dt'. \tag{4.45}$$

Hence, on substituting this in (43), we find that

$$D = \lim_{t\to\infty} \frac{1}{2t} \int_0^t dt_1 \int_0^t dt_2 \, \langle u(t_1)u(t_2)\rangle$$

$$= \lim_{t\to\infty} \frac{1}{t} \int_0^t dt_1 \int_0^{t-t_1} dt' \, \langle u(t_1)u(t_1 + t')\rangle, \tag{4.46}$$

so that, on changing the order of integration we obtain†

$$D = \int_0^\infty \langle u(t_0)u(t_0 + t)\rangle \, dt. \tag{4.47}$$

Equation (47), which arises quite naturally in the problem of Brownian motion, is a specific case of the so-called fluctuation–dissipation theorem, since D, which is related to the friction coefficient γ by Eq. (44), is associated with the dissipation of energy. In general, according to this theorem, the correlation function of the variable x can be expressed in terms of the solution to an entirely different problem, namely, the system's response to an external force f which leads to a deviation of the value of x from its equilibrium value. In our example, the fluctuations of u (or of x) were induced by the random force $f(t)$. Similarly, in the general case one can formally regard the thermal fluctuations in the value of x as being produced by the action on the system of a random external force f. From this point of view, the fluctuation–dissipation theorem expresses the requirement that the power entering the system as a result of the work done by external forces must be entirely dissipated and transferred to the thermal reservoir if the system's equilibrium

---

† In deriving (47) from (46), we have used the facts that $\langle u(t_1)u(t_1 + t')\rangle$ is independent of t_1, its derivative with respect to t' vanishes, and so does its limit as $t' \to \infty$.

state is not to be disturbed. The main importance of the fluctuation–dissipation theorem is that it relates the thermodynamic properties of a system (viz., fluctuations) to its transport coefficients.

Finally, we note that the fluctuation–dissipation theorem can affect the analytic properties of the transport coefficients. An interesting example of this is provided by the coefficient of viscosity η. It is found from hydrodynamics that the correlation function associated with η, analogous to Eq. (47), has a slowly decreasing tail for long times; for three dimensional systems this is proportional to $t^{-3/2}$, and for two-dimensional ones to t^{-1}. Hence, on performing the integration over t involved in the analog of (47), we find that for two-dimensional systems η has a logarithmic singularity. This extraordinary result, which makes no sense physically, shows that the theory is not self-consistent, i.e., that usual hydrodynamics "does not exist" in two dimensions.

4.2 Analytic Properties of the Scattering Matrix

Scattering Amplitudes and the S-Matrix

The standard quantum-mechanical approach to the description of a system's microscopic states consists of finding the relevant Hamiltonian operator and calculating its eigenvalues and eigenfunctions. However, these microscopic states are almost always studied experimentally by examining the interaction between the system and an incident particle scattered from it, and the system's properties are deduced from this scattering. In order to avoid the problems that arise in a full treatment of the interaction between the system and the scattering particle, it is advantageous and also natural to study the quantity known as the scattering matrix or S-matrix defined in the preceding section. This scattering matrix $S(k)$ determines the scattering cross section as a function of the wave number k of the incident particle. By making use of its analytical continuation for complex values of k, we can take into account the discrete values of the energy for bound states of the scattering system which is such an important feature of the solutions of Schrödinger's equation. In addition, we can also derive dispersion relations that provide a connection between different experimentally observable quantities.

If an incident spinless free particle moving along the z axis is scattered by some system situated at the origin, the asymptotic form of its wave function ψ can be expressed as

$$\psi(\mathbf{r}) = e^{ikz} + f(\theta)e^{ikr}/r, \qquad (4.48)$$

where θ is the angle between $\mathbf{r}$ and the z axis, and we assume for convenience, that the system has axial symmetry about this axis. Here, the first term describes the incident particle, and the second one the scattered particle, which at long distances from the origin can be represented by an outgoing spherical wave. All the information about the scattering process is contained in the function $f(\theta)$, which is called the scattering amplitude. This function determines the probability that the scattered particle will pass through a surface element $dS = r^2 \, d\Omega$, where $d\Omega$ is an element of solid angle,

$$d\Omega = 2\pi \sin \theta \, d\theta. \tag{4.49}$$

The scattering cross section $d\sigma$ is defined as the ratio of this probability to the current density of the incident wave, and so

$$d\sigma = |f(\theta)|^2 \, d\Omega. \tag{4.50}$$

The fact that $f(\theta)$ describes only the asymptotic behavior of the scattered particle and ignores the details of the interaction is in accord with the usual experimental situation. Moreover, according to relativistic quantum theory, it is in principle impossible to observe the behavior of a particle during the interaction because of the restrictions imposed by Heisenberg's uncertainty principle. It is in connection with the relativistic theory that the S-matrix approach has turned out to be the most fruitful.

We now restrict our attention to the simplest case of spherically symmetric scattering, in which the scattering amplitude f is independent of the angle of scatter θ. On writing $z = r \cos \theta$ and taking an average over angle in Eq. (48),† we find that if $f(\theta) = f$, a constant, then

$$\bar{\psi} = \frac{1}{4\pi} \int \psi \, d\Omega = \frac{\sin kr}{kr} + \frac{f}{r} e^{ikr} = \frac{i}{2kr} [e^{-ikr} - (1 + 2ikf)e^{ikr}]. \tag{4.51}$$

A comparison of this result with Eq. (9) shows that the spherically symmetric scattering amplitude f is related to the scattering matrix $S(k)$ defined in Section 4.1 according to

$$S(k) = 1 + 2ikf. \tag{4.52}$$

The unitarity requirement of Eq. (10), $S^*(k)S(k) = 1$, can now be expressed as a restriction on the function f, namely

$$|1 + 2ikf|^2 = 1 \tag{4.53}$$

Incidentally, Eq. (53) also follows from the requirement that a scatterer does not absorb or emit particles, so that the particle current density of Eq. (3.91) must vanish. Thus, in view of Eq. (3.92), the physical meaning of the unitarity

† An average over angle is taken so as to relate the present problem of an incident plane wave, the usual situation in practice, to the mathematically simpler problem of an incident spherical wave considered in Section 4.1.

condition for the scattering matrix S is that the probability density of particles must be conserved.

Since k is real, alternative forms of Eq. (53) are

$$\text{Im } f = k|f|^2 \quad \text{or} \quad \text{Im}(1/f) = \text{Im}(f^*/|f|^2) = -k. \quad (4.53a)$$

Thus, the scattering amplitude $f(k)$ can be written in the form

$$f(k) = 1/(g_0(k^2) - ik). \quad (4.54)$$

While the requirement of unitarity determines the imaginary part of $1/f$, its real part $g_0(k^2)$ depends on the detailed form of the scattering potential. Its form will be derived below for a specific case.

Analytical Properties of the S-Matrix

In order to study the analytical properties of the scattering amplitude and the S-matrix, we extend them to complex values of the wave vector k by means of analytic continuation. A case that has a particularly simple physical interpretation is that in which k is pure imaginary, $k = ik_1$. According to Eq. (7), the energy of a particle with imaginary k is real and negative. Hence, the bound states of a particle in a given scattering potential must be described by such a wave vector. In order that the wave function of Eq. (9) should correspond to a bound state, the term whose amplitude increases with increasing r must vanish. Hence, $S(k)$ must become zero at points on the negative imaginary axis corresponding to the energies of such states. In view of Eqs. (52) and (54), if $S(k) = 0$, then $g_0(k^2) - ik = -2ik$, and so $g_0(k^2) + ik = 0$. Thus, for each zero of $S(k)$ on the negative imaginary axis there is a pole at the corresponding point on the positive part of the imaginary axis. While these are the only poles of $S(k)$ in the upper half-plane, $S(k)$ does have an additional type of singularity in this region, which we consider shortly.

When $k = ik_1$, we can substitute from Eq. (54) into Eq. (51) and rewrite the latter

$$\bar{\psi} = \frac{1}{2k_1 r} \left[e^{k_1 r} - \frac{(g_0 - k_1)}{(g_0 + k_1)} e^{-k_1 r} \right]. \quad (4.55)$$

The existence of a pole in $S(k)$ for $k = ik_1 = -ig_0(-k_1^2)$ means that the scattering increases sharply if the energies of the incident particle and of a bound state of the scattering potential have the same absolute value. This condition for the increased scattering, which is known as resonance scattering, is found to be fulfilled only for slow particles, where the energy of the resonant bound state is small compared to the depth of the potential well. Finally, we note that we have restricted our attention to the simples case of the s-scattering ($l = 0$, i.e., spherically symmetric) of spinless particles.

Scattering by a Square Well Potential

As an example of the above theory, we consider now the scattering of a particle by the spherically symmetric square well potential

$$V(r) = \begin{cases} 0, & r > a, \\ -V_0, & r < a. \end{cases} \tag{4.56}$$

Outside the well, i.e., for $r > a$, the spherically symmetric wave function is of the form given by Eq. (51), so that

$$r\psi(r) = (1/k)\sin kr + fe^{ikr}. \tag{4.57}$$

The solution of Schrödinger's equation (with $l = 0$) for $r < a$, which is finite at $r = 0$, has a wave function ψ given by

$$r\psi(r) = A\sin \kappa r, \qquad \kappa = [2m(E + V_0)]^{1/2}/\hbar. \tag{4.58}$$

On matching ψ and its derivative at the boundary of the potential well, $r = a$, we find that

$$f(k) = \frac{\sin(ka) - (k/\kappa)\tan(\kappa a)\cos(ka)}{i(k^2/\kappa)\tan(\kappa a) - k}e^{-ika}. \tag{4.59}$$

Hence, from Eq. (52), the scattering matrix is given by

$$S(k) = \frac{1 + i(k/\kappa)\tan(\kappa a)}{1 - i(k/\kappa)\tan(\kappa a)}e^{-2ika}. \tag{4.60}$$

We now consider the cases of small and large wave vectors k of the incident particle. The terms small and large are relative to the two characteristic inverse lengths of the problem, $1/a$ and $(2mV_0)^{1/2}/\hbar$.

For a wave vector k that is much smaller than $1/a$, Eq. (59) becomes

$$f(k) = 1/(\kappa\cot(\kappa a) - ik), \tag{4.61}$$

which is just a special case of Eq. (54). In our case, $g_0(k^2) = \kappa\cot(\kappa a)$. Thus, it is an analytical function of k^2 or of E; if, in addition $k \ll (2mV_0)^{1/2}/\hbar$, $g_0(k^2)$ can be expanded in a power series in the small parameter E/V_0. A pole in $f(k)$ will appear for small k if $\cot(\kappa a)$ is close to zero, and so if κa is close to an odd multiple of $\pi/2$. This is in accordance with the discussion of resonance scattering following Eq. (55), since it is well known from quantum mechanics that $\kappa a \simeq (2n + 1)\pi/2$ is the condition for the existence of a spherically symmetric bound state with energy close to zero. If $ka \ll 1$ and $\kappa a \ll 1$ so that $V_0 \ll \hbar^2/ma^2$, the potential $V(r)$ can be treated as a perturbation in Schrödinger's equation. The Born theory of scattering can then be applied to $V(r)$, and the result obtained coincides, of course, with Eq. (161).

If the potential well is replaced by a potential barrier, i.e., if $-V_0$ is replaced by V_0 in Eq. (56), then $\cot(\kappa a)$ in Eq. (61) will be replaced by $\coth(\kappa a)$. In the limiting case of $\kappa a \ll 1, f \approx a$, and so the collision cross section, in view of Eq. (50), is just $\sigma = 4\pi a^2$. This case corresponds to an impenetrable sphere, for which the collision cross section according to classical mechanics is only a quarter of the above.

*Dispersion Relations

We now return to the problem of the square well potential, and consider the second limiting case, namely, that of large values of $|k|$ where k lies in the upper half of the complex plane. The factor e^{-2ika}, which appears in Eq. (60) for $S(k)$, has an essential singularity at $i\infty$, since as $k_1 \to \infty$, $e^{2k_1 a} \to \infty$ faster than any power of k_1. Thus, in order to study a function without an essential singularity at infinity, we must consider $S(k)e^{2ika}$. However, even this function is not suitable for the application of the Kramers–Kronig dispersion relations discussed in Section 4.1, since these can only be applied to a function that tends to zero (or some finite limit) as $|k| \to \infty$. The function $S(k)e^{2ika}$, on the other hand, is found not to behave like this,† and only $S(k)e^{2ika}/k$ has no singularity at infinity, and in fact vanishes there. In order to obtain a function that is also not singular at the origin we consider

$$F(k) = [S(k)e^{2iak} - S(0)]/k. \qquad (4.62)$$

If the scattering potential does not have any bound states, $F(k)$ is analytic in the upper half-plane and regular at $k = 0$, but does not satisfy the parity relations of Eq. (28). However, if we write

$$S(k)e^{2iak}/(2iak) = \phi'(k) + i\phi''(k), \qquad (4.63)$$

and use the fact that $S(0)$ is real according to Eq. (52), we find that, by analogy with Eqs. (34) and (35),

$$\phi''(k) = -\frac{S(0)}{2k} + \frac{2}{\pi}\int_0^\infty \frac{k\phi'(k')}{k^2 - k'^2}\,dk', \qquad (4.64)$$

$$\phi'(k) = \frac{2}{\pi}\int_0^\infty \frac{k'\phi''(k')}{k'^2 - k^2}\,dk'. \qquad (4.65)$$

If the scattering potential has bound states, these lead to the appearance of poles in $S(k)$ in the upper half of the complex plane, as we saw earlier. In

---

† This result is not immediately obvious, but can be obtained from experimental studies of collision cross sections at high energies, or from theoretical results such as Eq. (60).

such a case, we must use Cauchy's residue theorem, Eq. (11), in performing the integrals that lead to Eqs. (64) and (65). If we denote by σ_j the residue of the function $kS(k)e^{2iak}$ at the point k_j, we must then add to the right-hand sides of Eqs. (64) and (65) the expressions $\sum_j k\sigma_j/(k_j^4 - k_j^2 k^2)$ and $-i\sum_j \sigma_j/(k_j k^2 - k_j^3)$, respectively, where the sum is over all the bound states. Dispersion relations for the scattering matrix are widely used in elementary particle physics and the theory of reactions in nuclei, but the consideration of these topics is outside the scope of this book.

4.3   Dispersion Relations for Macroscopic Systems

In many physical problems, we are interested in finding or making use of the response of a macroscopic system to a given external stimulus. This stimulus can be regarded as the input to the system, and the response as the output that it produces. If the input is sufficiently small, the output is frequently a linear function of it, and we will restrict our attention to such systems. One example of such a problem is the electrical displacement **D** produced by an electric field **E**, which we considered in Section 4.1. Other examples include the current produced in a circuit by an applied voltage, the waves or particles scattered by a system as a function of the incident beam, and the output produced in a radiation detector system or servomechanism by the appropriate input. For all these linear systems, the relationship between the input $q(t)$ and the output $p(t)$ can be described by a generalized susceptibility $Y(\omega)$, which relates their Fourier transforms according to Eq. (25), $P(\omega) = Y(\omega)Q(\omega)$. The fact that the output is caused by the input, and so cannot precede it, can then lead to the Kramers–Kronig relations, Eqs. (31) and (32) or (34) and (35), between the real and imaginary parts of $Y(\omega)$. However, the validity of these relations in any given case depends on the asymptotic behavior of $Y(\omega)$ as $\omega \to \infty$, and we start by considering this restriction.

Convergence Conditions

The central theorem on which most discussions of causality and dispersion relations is based, known as Titchmarch's theorem, can be stated as follows. Let $F(\omega)$ be a square integrable function on the real ω axis, i.e., such that $\int_{-\infty}^{\infty} |F(\omega)|^2 \, d\omega$ exists. Then if $F(\omega)$ is the Fourier transform of a square integrable function $f(t)$ which vanishes for $t < 0$, the real and imaginary parts of $F(\omega)$, which we denote by $F'(\omega)$ and $F''(\omega)$, respectively, are related by

$$F'(\omega) = \frac{1}{\pi} \int_{-\infty}^{\infty} \frac{F''(x)}{x - \omega} \, dx, \qquad (4.66a)$$

$$F''(\omega) = -\frac{1}{\pi} \int_{-\infty}^{\infty} \frac{F'(x)}{x - \omega} \, dx, \qquad (4.66b)$$

where $\fint$ denotes the principal part of the integral. Conversely, if $F(\omega)$ is square integrable and Eqs. (66) hold, $F(\omega)$ is the Fourier transform of a square integrable function $f(t)$ that vanishes for $t < 0$. Functions F' and F'' which are related according to Eqs. (66) are said to be Hilbert transforms of each other; extensive tables of such transforms have been published. Equations (66) are just Eqs. (31) and (32) that we derived in Section 4.1, with $F(\omega)$ in place of $Y(\omega) - y_0$.

The requirement that $F(\omega)$ be square integrable can be replaced by the requirement that $f(t)$ be bounded and square integrable. This is the mathematical reason why we separated out the term $y_0 \, \delta(t)$, which is not bounded, in Eq. (24), and considered $Y(\omega) - y_0$ rather than $Y(\omega)$ in the subsequent analysis. If $y_0 = 0$, a sufficient condition for $Y(\omega)$ to be square integrable and the Kramers–Kronig relations to apply is that the output $p(t)$ be square integrable whenever the input $q(t)$ is. Such a condition will be obeyed if, as often happens in scattering problems, $\int_{-\infty}^{\infty} |q(t)|^2 \, dt$ and $\int_{-\infty}^{\infty} |p(t)|^2 \, dt$ are proportional to the total energy of the input and output signals. The proof of this theorem is not simple, and can be found in standard textbooks on Fourier transforms or dispersion relations. We just remark that the contour used to derive Eqs. (66) is not the infinite semicircle used in Section 4.1, but rather an infinite rectangle in the upper half-plane, with one of its sides along the real axis.

As an example of the importance of the convergence requirements, let us consider the circuit shown in Fig. 4. This contains a generator that provides a potential $v(t)$, a resistance R, and a solenoid of self-inductance L, while the mutual inductance between the solenoid and its core will be denoted by

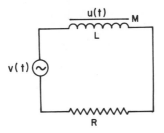

Fig. 4. An electrical circuit with resistance and self-inductance.

M. Both the current $i_1(t)$ in the circuit and the emf $u(t)$ generated in the core are caused by and depend linearly on the external potential $v(t)$. Hence, we can write, by analogy with Eq. (25),

$$I_1(\omega) = Y(\omega)V(\omega), \qquad (4.67)$$

$$U(\omega) = C(\omega)V(\omega). \qquad (4.68)$$

Here, as usual, a function denoted by a capital letter is the Fourier transform of the one denoted by the corresponding lowercase letter. If the core is electrically conducting, the induced emf $u(t)$ will produce currents in it, known as eddy currents, even though the core is not part of a closed circuit. Provided that these eddy currents are small, the current $i_1(t)$ satisfies approximately the equation

$$L\, di_1/dt + Ri_1 = v(t), \qquad (4.69)$$

while

$$u(t) = M\, di_1/dt. \qquad (4.70)$$

We then readily find that

$$Y(\omega) = 1/(R - i\omega L), \qquad (4.71)$$

$$C(\omega) = -Mi\omega/(R - i\omega L). \qquad (4.72)$$

Since $Y(\omega)$ is a square integrable function of ω, its real and imaginary parts are related by Eqs. (31) and (32) or (34) and (35), with $y_0 = 0$ and $A = 0$. On the other hand, $|C(\omega)| \to M/L$ as $|\omega| \to \infty$, so that $C(\omega)$ as it stands is not square integrable. In this particular case, we can write

$$C(\omega) - M/L = -(MR/L)/(R - i\omega L), \qquad (4.73)$$

so that the Kramers–Kronig relations apply to $C(\omega) - M/L$. However, this sort of result will not occur in all problems, and so one cannot assume blindly that these relations apply whenever there is a linear causal relationship between two quantities. Generalizations of the dispersion relations are possible for systems in which $Y(\omega)$ is not square integrable, but their form is much more complicated.

The original derivation of the Kramers–Kronig relations was for $\tilde{n}(\omega) - 1$, where $\tilde{n}(\omega)$ is the complex refractive index of a medium for light of angular frequency ω. At first sight, it is not obvious why these relations should apply to $\tilde{n} - 1$, since it does not seem to describe a causal process. However, the velocity of light, and hence the refractive index, can be derived not only from Maxwell's equations with a macroscopic dielectric constant but also in terms of molecular scattering. In this latter approach, the velocity of light in a medium differs from its value in vacuum because of the interaction between the incident waves and the waves scattered from the molecules in the medium. In particular, it is found that $\tilde{n} - 1$ is proportional to the amplitude of the wave scattered by a thin slab of the medium. Since the scattered wave is caused by and depends linearly on the incident wave, $\tilde{n} - 1$ can be regarded as describing a causal process, to which our previous analysis applies.

Applications of Dispersion Relations

One important application of dispersion relations is to check the consistency of experimental observations and of the analytical formulas to which they are fitted. This is especially useful if the data contains large random errors and can be represented approximately by more than one analytical expression.

One example of such a consistency check is provided by the complex dielectric constant $\varepsilon(\omega)$. For many dielectric materials, it is found that $\varepsilon''(\omega)$ is proportional to ω^{n-1} for a wide range of frequencies, where $0 < n < 1$. An examination of tables of integral transforms shows that the Hilbert transform of such a function is itself proportional to ω^{n-1}. Thus, since $\varepsilon' - 1$ and ε'' are related by the Kramers–Kronig relations, the ratio $\varepsilon''(\omega)/[\varepsilon'(\omega) - 1]$ should be constant, and this is found to be the case. In fact, though, since $\varepsilon''(\omega)$ must tend to zero as ω tends to infinity, this form for $\varepsilon''(\omega)$ can only apply to a finite frequency range. Fortunately, this does not matter too much, since the integrals for $\varepsilon'(\omega)$ and $\varepsilon''(\omega)$ are dominated by the behavior of the integrand close to ω. As a result, the deviations of $\varepsilon''(\omega)$ from proportionality to ω^{n-1} will only affect the value of $\varepsilon'(\omega) - 1$ for large frequencies ω.

An interesting and much less obvious use of dispersion relations occurs in connection with the reflection of light from a system containing several parallel layers of different materials. For the sake of simplicity, we consider only the case of a beam of light incident normally from the air onto a single layer of material, of thickness h, that rests on a semi-infinite substrate, as shown in Fig. 5. Let $\tilde{n}_1(\omega)$, $\tilde{n}_2(\omega)$, and $\tilde{n}_3(\omega)$ be the complex refractive indices of air, the layer, and the substrate, respectively. The values of $\tilde{n}_1(\omega)$ and $\tilde{n}_3(\omega)$ are known, and the aim of the experiments is to derive $\tilde{n}_2(\omega)$ from measurements of the reflectance of the system $r_s(\omega)$, which is the ratio of the electric fields of the reflected and incident waves at the air/sample interface. It can readily be shown that $r_s(\omega)$ is related to the $\tilde{n}_j(\omega)$ by the formula

$$r_s(\omega) = (u_{12} + u_{23}e^{2i\beta})/(1 + u_{12}u_{23}e^{2i\beta}), \qquad (4.74)$$

where

$$u_{jk} = (\tilde{n}_j - \tilde{n}_k)/(\tilde{n}_j + \tilde{n}_k), \qquad \beta = \tilde{n}_2\omega h/c, \qquad (4.75)$$

and c is the velocity of light in vacuum.

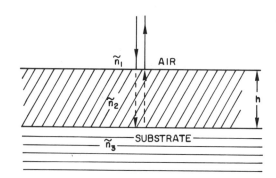

Fig. 5.   The reflection of light from a layer of material on a substrate.

If we write

$$r_s(\omega) = \rho_s \exp(i\theta_s), \tag{4.76}$$

the amplitude ρ_s is just the square root of the ratio of the intensities of the incident and reflected waves, and so can readily be measured. However, the change of phase θ_s is much more difficult to measure, and so we would like to find dispersion relations that express θ_s in terms of ρ_s.

The fact that r_s describes a causal process, so that its real and imaginary parts are connected by the Kramers–Kronig relations, only leads to a complicated integral equation for θ_s. Instead, we start by defining a model substance whose refractive index $\tilde{n}_b(\omega)$ has the same asymptotic dependence on ω for large ω as $\tilde{n}_2(\omega)$, but which has no absorption bands in the frequency range that we measure, so that $\tilde{n}_b(\omega)$ is a smooth function of ω in that range. Let

$$r_b(\omega) = \rho_b \exp(i\theta_b) \tag{4.77}$$

be the reflectance of the system when the layer of material of refractive index $\tilde{n}_2$ is replaced by a layer of the model substance. Then, as we now show, the real and imaginary parts of

$$G(\omega) = \ln[r_s(\omega)/r_b(\omega)] \tag{4.78}$$

are Hilbert transforms of each other, and $G(\omega)$ satisfies the conditions of Titchmarch's theorem.

The usual dielectric constants $\varepsilon_j(\omega) = \tilde{n}_j^2(\omega)$ of the various media can be extended analytically into the upper half of the complex ω plane, and, as we saw in Section 4.1, are nonzero analytic functions of ω on and above the real ω axis. Hence, the $\tilde{n}_j(\omega)$ have similar properties, and so, in view of Eq. (74), the analytic continuation of $r_s(\omega)/r_b(\omega)$ will also be a nonzero analytic function of ω in this region, provided that $r_b(\omega)$ is chosen in such a way that

r_s/r_b never becomes zero or infinite. As $|\omega| \to \infty$, the real and model systems become identical, so that $r_s(\omega)/r_b(\omega) \to 1$. Hence, $G(\omega)$ is an analytic function of ω on and above the real ω axis, and $G(\omega) \to 0$ as $|\omega| \to \infty$. We now proceed as in Section 4.1, and calculate the integral of

$$f(z) = G(z)/(z - \omega) \tag{4.79}$$

round the contour shown in Fig. 3. Along the semicircle C_R, which can be described by $z = Re^{i\phi}, 0 \le \phi \le \pi$, we find that

$$\left| \int_{C_R} f(z)\, dz \right| \le \int_0^\pi |f(z)| \,|iRe^{i\phi}|\, d\phi = \int_0^\pi \left| \frac{R}{Re^{i\phi} - \omega} \right| |G(Re^{i\phi})|\, d\phi, \tag{4.80}$$

and this tends to zero as $R \to \infty$ since $|R/(Re^{i\phi} - \omega)| \to 1$ and $|G(Re^{i\phi})| \to 0$. Thus, just as in Eq. (30),

$$\int_{-\infty}^{\infty} \frac{G(z)\, dz}{z - \omega} = i\pi G(\omega), \tag{4.81}$$

so that the real and imaginary parts of $G(\omega)$ are Hilbert transforms of each other, just as in Eqs. (31) and (32), In particular,

$$\theta_s - \theta_b = -\frac{1}{\pi} \int_{-\infty}^{\infty} \frac{\ln[\rho_s(x)/\rho_b(x)]}{x - \omega}\, dx, \tag{4.82}$$

which is just the sort of equation for θ_s that we wanted.

We now show that $G(\omega)$ is square integrable. As $\omega \to \infty$, $\tilde{n}_j(\omega) \to 1$ and hence $r_s(\omega) \to 0$, $r_b(\omega) \to 0$. Thus, r_s and r_b are analytic functions of $1/\omega$ as $\omega \to \infty$, and since they have the same asymptotic behavior, we can write

$$r_s(\omega)/r_b(\omega) \sim 1 + A/\omega^k, \qquad k = \text{positive integer.} \tag{4.83}$$

Hence, $G(\omega) \sim A/\omega^k$ as $\omega \to \infty$, and this is a sufficient condition for $G(\omega)$ to be square integrable. It follows from Titchmarch's theorem that $G(\omega)$ is a causal transform, i.e., the Fourier transform of a function $g(t)$ that vanishes for $t < 0$. It is far from obvious what sort of input and output are related by $G(\omega)$, and this example shows how the ideas of causal functions and dispersion relations can be applied to problems that do not involve just simple inputs and outputs.

Quantum-Mechanical Approach

In order to extend the use of dispersion relations to quantum-mechanical systems, it is desirable, when possible, to express the cause and effect, or input and output, in terms of the elements of the system's Hamiltonian. To do this, we assume that an external influence or perturbing force $q(t)$ gives rise to a perturbing term in the system's Hamiltonian of the form

$$\hat{v} = -\hat{p}q(t). \tag{4.84}$$

We choose $p(t)$, the expectation value of $\hat{p}$, in such a way that it vanishes when the system is in its equilibrium state in the absence of the perturbation. In the presence of the perturbation, we assume that we can write

$$p(t) = y_0 q(t) + \int_0^\infty y(\tau)q(t - \tau)\, d\tau, \tag{4.85}$$

which is exactly the same as Eq. (24), and the analysis then proceeds as in Section 4.1.

One immediate consequence of our definition of $p(t)$, $y(t)$, and $Y(\omega)$ in terms of an element of the Hamiltonian concerns the rate of dissipation of energy. A real force

$$q(t) = \tfrac{1}{2}(q_0 e^{-i\omega t} + q_0^* e^{i\omega t}) \tag{4.86}$$

will, according to Eq. (85), produce a response

$$p(t) = \tfrac{1}{2}[q_0 Y(\omega)e^{-i\omega t} + q_0^* Y_0^*(\omega)e^{i\omega t}], \tag{4.87}$$

since $Y(-\omega) = Y^*(\omega)$ according to Eq. (28). The change in the system produced by the force $q(t)$ is accompanied by the absorption of energy (which is subsequently dissipated as heat) at a rate

$$w(t) = dE/dt, \tag{4.88}$$

where $E(t)$ is the mean energy of the system. It is a well-known quantum-mechanical result that dE/dt equals the expectation value of $\partial \hat{H}/\partial t$. Since, in our case, the only term in $\hat{H}$ that depends explicitly on time is $\hat{v}$, we find, from Eq. (84) that

$$w(t) = \text{expectation value of}\quad [-\hat{p}\, dq/dt]. \tag{4.89}$$

On substituting in this formula the values of $p(t)$ and $q(t)$, given by Eqs. (86) and (87), and averaging over a period of $q(t)$, we readily find that

$$W(\omega) = (i\omega/4)[Y^*(\omega) - Y(\omega)]|q_0|^2 = \tfrac{1}{2}\omega Y''(\omega)|q_0|^2. \tag{4.90}$$

Thus, the dissipation of energy is determined by the imaginary part of the generalized susceptibility $Y(\omega)$. Since any real process is always accompanied by some dissipation of energy, W and so $Y''(\omega)$ must be positive for all finite positive frequencies.

Calculation of the Dielectric Constant

A quantum-mechanical calculation of the rate of absorption of energy as the result of a perturbation is much simpler than one of the expectation value of the output that it produces. Thus, the easiest way to obtain quantum-mechanical formulas for a generalized susceptibility $Y(\omega)$ is to use Eq. (90), which relates the rate of absorption of energy to $Y''(\omega)$, and the Kramers–Kronig relations to obtain $Y'(\omega)$ from $Y''(\omega)$. As an example of this procedure, we consider the dielectric constant of an insulating medium, in which the electrons can each be associated with a definite atom (or molecule). We treat the electromagnetic fields classically, for the sake of simplicity, rather than use a full quantum electrodynamical approach.

In the analysis of the absorption of energy from electromagnetic waves, it is most convenient to use the vector potential $A(r, t)$, and choose a gauge (see Sections 3.1 and 3.2) in which the scalar potential vanishes. In classical mechanics, the effect of such waves on a particle's Hamiltonian is to replace the kinetic energy term $p^2/2m$, where $\mathbf{p}$ is the particle's momentum, by a term $(\mathbf{p} + e\mathbf{A}/c)^2/2m$ for an electron of charge $-e$, as in Eq. (3.94). Hence, to first order in the fields, the waves introduce into the Hamiltonian a perturbation term

$$\hat{v}(\mathbf{r}, t) = (e/mc)\mathbf{A} \cdot \hat{\mathbf{p}}, \tag{4.91}$$

where $\hat{\mathbf{p}}$ is the momentum operator. It is convenient to choose the x axis in the direction of $\mathbf{A}$ for linearly polarized waves. For these we can use the Fourier representation

$$A(\mathbf{r}, t) = \int_{-\infty}^{\infty} A(\omega)e^{i(\mathbf{k}\cdot\mathbf{r} - \omega t)}\, d\omega. \tag{4.92}$$

According to Fermi's golden rule, a perturbation $\hat{v}$ of frequency ω induces transitions from a state j, of energy E_j and wave function u_j, to a state k, having energy E_k and wave function u_k, at a rate

$$M_{jk} = (2\pi/\hbar^2)|\langle u_k|\hat{v}|u_j\rangle|^2\delta(\omega - \omega_{kj}), \tag{4.93}$$

where

$$\hbar\omega_{kj} = E_k - E_j. \tag{4.94}$$

In our case,

$$|\langle u_k|\hat{v}|u_j\rangle|^2 = (e^2/m^2c^2)|A(\omega_{jk})|^2|\langle u_k|e^{i\mathbf{k}\cdot\mathbf{r}}\hat{p}_x|u_j\rangle|^2. \tag{4.95}$$

For an atom, whose dimensions are much smaller than the wavelength of light, we can write $e^{i\mathbf{k}\cdot\mathbf{r}} \approx 1$ in Eq. (95). It is convenient to make use of the identity

$$\langle u_k|\hat{p}_x|u_j\rangle = m(d/dt)\langle u_k|\hat{x}|u_j\rangle = -im\omega_{kj}x_{kj}, \tag{4.96}$$

where we have written x_{kj} for $\langle u_k | \hat{x} | u_j \rangle$. On combining Eqs. (93), (95), and (96), we find that the transition rate can be expressed as

$$M_{jk} = (2\pi e^2/\hbar^2 c^2)\omega^2 |A(\omega_{jk})|^2 |x_{kj}|^2 \delta(\omega - \omega_{kj}). \qquad (4.97)$$

Hence, the total rate of absorption of energy from the field by an electron in state j, if it can only make transitions to discrete states k, is just

$$W = \int d\omega \sum_k \hbar\omega_{kj} M_{jk} = \frac{2\pi e^2}{\hbar c^2} \int_0^\infty d\omega \sum_k \omega^3 |A(\omega)|^2 |x_{kj}|^2 \delta(\omega - \omega_{kj}). \qquad (4.98)$$

Since the electric field $E(\omega)$ associated with the vector potential $A(\omega)$ is of amplitude $|\omega A(\omega)|/c$, we can rewrite Eq. (98) in the form

$$W = \frac{2\pi e^2}{\hbar} \int_0^\infty \omega |E(\omega)|^2 \sum_k |x_{kj}|^2 \delta(\omega - \omega_{kj}) \, d\omega. \qquad (4.99)$$

We now wish to derive a formula for $\alpha''(\omega)$, the polarizability per electron, by comparing Eq. (99) with the integral over frequencies of Eq. (90), substituting α for Y. However, the perturbing force considered in deriving Eq. (90) was of the form (86), while we are now considering an electric field with components

$$\mathbf{E}(\mathbf{r}, t) = \mathbf{E}(\omega) \exp[i(\mathbf{k} \cdot \mathbf{r} - \omega t)] + \mathbf{E}^*(\omega) \exp[-i(\mathbf{k} \cdot \mathbf{r} - \omega t)]. \qquad (4.100)$$

Hence, we must replace $|q_0|^2$ by $4|E(\omega)|^2$ in Eq. (90), so that on integrating that equation we find

$$W = 2 \int_0^\infty \omega \alpha''(\omega) |\mathbf{E}(\omega)|^2 \, d\omega. \qquad (4.101)$$

On comparing Eqs. (99) and (101), we see that

$$\alpha''(\omega) = \frac{\pi e^2}{\hbar} \sum_k |x_{kj}|^2 \delta(\omega - \omega_{kj}). \qquad (4.102)$$

For a system containing N electrons per unit volume, it follows from Eq. (17) that

$$\varepsilon''(\omega) = 4\pi N \alpha''(\omega). \qquad (4.103)$$

Hence, we obtain the quantum-mechanical formula

$$\varepsilon''(\omega) = \frac{4\pi^2 e^2 N}{\hbar} \sum_k |x_{kj}|^2 \delta(\omega - \omega_{kj}), \qquad (4.104)$$

which can be compared with Eq. (37).

Oscillator Strengths and Quantum-Mechanical Sum Rules

It is conventional to define the oscillator strength per electron, $f(\omega)$, of a medium containing N electrons per unit volume, by the equation

$$f(\omega) = (m/2\pi^2 e^2 N)\omega\varepsilon''(\omega). \tag{4.105}$$

A comparison of this equation with Eq. (34) then shows that

$$\varepsilon'(\omega) - 1 = \frac{4\pi e^2 N}{m} \int_0^\infty \frac{f(x)\,dx}{x^2 - \omega^2}. \tag{4.106}$$

On comparing Eqs. (105) and (106) with Eq. (104) we see that

$$f(\omega) = \sum_k f_{kj}\,\delta(\omega - \omega_{kj}), \tag{4.107}$$

where

$$f_{kj} = (2m\omega_{kj}/\hbar)|x_{kj}|^2 \tag{4.108}$$

is the oscillator strength associated with the transition from state j to state k. From Eqs. (106) and (107), we see that as $\omega \to \infty$ [provided that $f(\omega)$ then tends rapidly to zero as we expect]

$$\varepsilon'(\omega) - 1 \sim -\frac{4\pi e^2 N}{m\omega^2} \sum_k f_{kj}. \tag{4.109}$$

Thus, a sum rule for the oscillator strengths f_{kj} will enable us to derive a quantum-mechanical sum rule analogous to Eq. (38).

In order to derive this sum rule, we note that, in view of Eqs. (96) and (108),

$$f_{kj} = (2i/\hbar)(p_x)_{jk}x_{kj} = (-2i/\hbar)x_{jk}(p_x)_{kj} = (i/\hbar)[(p_x)_{jk}x_{kj} - x_{jk}(p_x)_{kj}]. \tag{4.110}$$

Now if the states k form a complete set of eigenstates, then for any pair of operators $\hat{A}$ and $\hat{B}$

$$(\hat{A}\hat{B})_{jj} = \sum_k A_{jk} B_{kj}. \tag{4.111}$$

Hence

$$\sum_k f_{kj} = (i/\hbar)[\hat{p}_x, \hat{x}]_{jj} = 1, \tag{4.112}$$

in view of the usual commutation relations for x and p_x, since we are considering one-electron wave functions. Equation (112) is known as the Thomas–Reike–Kuhn sum rule. Thus, the quantum-mechanical formula (109) for the asymptotic behavior of $\varepsilon'(\omega)$ is identical with the classical one derived in Section 4.1,

$$\varepsilon'(\omega) \sim 1 - 4\pi N e^2/m\omega^2 = 1 - \omega_p^2/\omega^2, \tag{4.113}$$

say, where

$$\omega_{\mathrm{p}} = (4\pi N e^2/m)^{1/2} \tag{4.114}$$

is the plasma frequency of the medium.

In addition, from Eqs. (105), (107), and (112), we see that

$$\int_0^\infty \omega\varepsilon''(\omega)\, d\omega = \frac{2\pi^2 e^2 N}{m} \sum_k f_{kj} = \frac{\pi\omega_{\mathrm{p}}^2}{2}, \tag{4.115}$$

which is identical to the sum rule (36) that we derived in Section 4.1. From this sum rule, we see that $\omega\varepsilon''(\omega) \to 0$, as $\omega \to \infty$, more rapidly than ω^{-1}. Hence, since $\varepsilon''(\omega)$ is an analytic odd function of $1/\omega$ for large ω it must be $O(\omega^{-3})$. Thus, for insulators

$$\varepsilon(\omega) \sim 1 - \omega_{\mathrm{p}}^2/\omega^2 + O(\omega^{-3}). \tag{4.116}$$

Since the complex refractive index $\tilde{n}(\omega)$ is defined by

$$\tilde{n}^2(\omega) = \varepsilon(\omega), \tag{4.117}$$

it follows that for insulators, as $\omega \to \infty$,

$$\tilde{n}(\omega) \sim 1 - \tfrac{1}{2}\omega_{\mathrm{p}}^2/\omega^2 + O(\omega^{-3}). \tag{4.118}$$

Additional Sum Rules

The sum rule for $\varepsilon''(\omega)$ of Eq. (115) is just one example of a whole class of sum rules for the generalized susceptibility $Y(\omega)$. Quite generally, if

$$Y(\omega) - y_0 \sim -B/\omega^2, \quad \text{as} \quad \omega \to \infty, \tag{4.119}$$

where y_0 and B are real, then $xY''(x) = O(x^{-2})$ as $x \to \infty$. Hence, for large ω, we can neglect x in comparison with ω in the denominator of the integrand of Eq. (34), for all values of x that contribute appreciably to the integral. Thus, as $\omega \to \infty$, we find from Eq. (34) that

$$Y'(\omega) - y_0 \sim -\frac{2}{\pi\omega^2} \int_0^\infty xY''(x)\, dx. \tag{4.120}$$

A comparison of Eqs. (119) and (120) then shows that

$$\int_0^\infty \omega Y''(\omega)\, d\omega = \frac{\pi B}{2}. \tag{4.121}$$

Thus, for instance, if we write the complex refractive index as

$$\tilde{n}(\omega) = n(\omega) + ik(\omega), \tag{4.122}$$

then from Eqs. (118) and (121) we find that, for an insulating medium,

$$\int_0^\infty \omega k(\omega)\, d\omega = \frac{\pi \omega_p^2}{4}. \tag{4.123}$$

Sum rules such as those of Eqs. (115) and (123) are of great practical value, since they enable us to check whether measurements over a finite frequency range have encompassed all the system's absorption peaks [which are associated with $Y''(\omega)$ according to Eq. (90)]. However, various other types of sum rules exist which can be useful for analyzing experimental results or checking their consistency, and so we now describe briefly two fairly general techniques for deriving such sum rules.

In the first of these techniques, we integrate both sides of the Kramers–Kronig relations (34) over ω from $\omega = 0$ to $\omega = \infty$, and change the order of integration on the right-hand side. In order to do this, we note that if $g(x)$ is an arbitrary continuous function of x and $xg(x) \to 0$ as $x \to \infty$, then

$$\int_0^\infty d\omega \int_0^\infty dx\, \frac{g(x)}{x^2 - \omega^2} = \int_0^\infty dx\, g(x) \int_0^\infty \frac{d\omega}{x^2 - \omega^2}, \tag{4.124}$$

and, in accordance with one of the definitions of the δ-function,

$$\int_0^\infty \frac{d\omega}{x^2 - \omega^2} = -\frac{\pi^2}{4}\, \delta(x). \tag{4.125}$$

Hence,

$$\int_0^\infty d\omega \int_0^\infty dx\, \frac{g(x)}{x^2 - \omega^2} = -\frac{\pi^2}{4} \lim_{x \to 0} g(x). \tag{4.126}$$

On applying this technique to Eq. (34), and using Eq. (33), we find that

$$\int_0^\infty (Y'(\omega) - y_0)\, d\omega = -\frac{\pi}{2} \lim_{\omega \to 0} (\omega Y''(\omega)) = -\frac{\pi A}{2}. \tag{4.127}$$

For a medium in which $A = 0$, we can also apply this method to $Y''(\omega)/\omega$, using Eq. (35), and so obtain

$$\int_0^\infty \frac{Y''(\omega)}{\omega}\, d\omega = \frac{\pi}{2} (Y'(0) - y_0). \tag{4.128}$$

Both these equations apply to an insulating medium with $Y(\omega) = \varepsilon(\omega)$ or $\tilde{n}(\omega)$ and $y_0 = 1$, while Eq. (127) also applies to $\varepsilon(\omega)$ for a conducting medium, with A related to the dc conductivity σ, according to Eq. (23), by

$$A = 4\pi\sigma. \tag{4.129}$$

A similar procedure, though rather more complicated, can be used to derive sum rules for powers of $Y'(\omega)$ and $Y''(\omega)$. These powers are expressed as the products of integrals by means of the Kramers–Kronig relations, and then integrated over ω from 0 to ∞, with an appropriate change in the order of integration of the multiple integrals to permit their evaluation.

Another technique for deriving additional sum rules is based on the use of convolutions of $y(t)$. We define

$$y_m(t) = \int_{-\infty}^{\infty} y_{m-1}(t - \tau)y(\tau)\, d\tau, \quad m > 1; \qquad y_1(t) = y(t). \quad (4.130)$$

Since $y(t) = 0$ for $t < 0$, we readily find by induction that $y_m(t) = 0$ for $t < 0$. From the convolution theorem, the Fourier transform of y_m is the product of that of y_{m-1} and y. Hence, since $Y(\omega) - y_0$ is the Fourier transform of y, we see that

$$y_m(t) = \frac{1}{2\pi} \int_{-\infty}^{\infty} [Y(\omega) - y_0]^m e^{-i\omega t}\, d\omega. \quad (4.131)$$

Moreover, if $Y(\omega) - y_0$ is square integrable and has no singularities on the real axis, then $(Y(\omega) - y_0)^m$ is also square integrable. Hence, according to Titchmarch's theorem, the real and imaginary parts of $(Y(\omega) - y_0)^m$ are connected by the Kramers–Kronig relations. This result, and the sum rules that can be derived from it, can be used to provide additional checks for the consistency of experimental observations or the formulas postulated to describe them. An advantage of this result is that the larger the value of m, the faster $(Y(\omega) - y_0)^m$ tends to zero as ω tends to infinity, so that measurements over a finite frequency range can be extrapolated with a much smaller error. We note that once again we have a causal transform for a function, in this case $y_m(t)$, which does not relate any simple input and output.

The Physical Meaning of Sum Rules and Dispersion Relations

Sum rules such as those of Eqs. (115) and (127) are not just useful mathematical formulas. Rather, they express (or result from) properties of the response function $y(t)$ which leads to the generalized susceptibility $Y(\omega)$. For instance, in analyzing the dielectric constant of a system we write

$$D(t) = E(t) + 4\pi P(t), \quad (4.132)$$

where the relation of the polarization $P(t)$ to the field $E(t)$ is described by the response function $g(t)$, defined by

$$P(t) = \int_0^{\infty} g(t - \tau)E(\tau)\, d\tau. \quad (4.133)$$

Hence, in view of Eq. (17) and by analogy with Eqs. (24) and (26),

$$\varepsilon(\omega) - 1 = 4\pi \int_0^\infty g(t)e^{i\omega t}\, dt. \tag{4.134}$$

The inverse Fourier transform of this then shows us that

$$g(t) = 2 \int_{-\infty}^\infty (\varepsilon(\omega) - 1)e^{-i\omega t}\, d\omega. \tag{4.135}$$

Since ε' and ε'' are, respectively, even and odd functions of ω,

$$g(0) = 4 \int_0^\infty (\varepsilon'(\omega) - 1)\, d\omega = -8\pi^2\sigma(0), \cdot \tag{4.136}$$

in view of Eqs. (127) and (129). Thus, the sum rule (127) in this case expresses the fact that only the conduction current, and not the polarization, can respond instantaneously to an applied electric field.

In addition, if in Eq. (135) we write $\omega - i\delta$ in place of ω, where δ is a small positive quantity, we see that for $t > 0$

$$g(t) = \lim_{\delta \to 0} 2 \int_{-\infty}^\infty (\varepsilon(\omega) - 1)e^{-i\omega t}e^{-\delta t}\, d\omega, \tag{4.137}$$

and hence that its derivative

$$\dot{g}(t) = \lim_{\delta \to 0} 2 \int_{-\infty}^\infty (-i\omega)(\varepsilon(\omega) - 1)e^{-i\omega t}e^{-\delta t}\, d\omega. \tag{4.138}$$

Thus, on letting t tend to zero from above, we see that for an insulator

$$\dot{g}(0+) = 4 \int_0^\infty \omega\varepsilon''(\omega)\, d\omega = 2\pi\omega_p^2, \tag{4.139}$$

in view of Eq. (115). Thus, this sum rule describes the derivative of the response function just after the field is applied.

The above results are typical examples of the fact that dispersion in the frequency domain results from the delayed response, often associated with the effects of inertia, in the time domain. In general, as we can see from Eq. (134), the value of $\varepsilon(\omega)$ at a given frequency ω is dominated by the behavior of $g(t)$ at times up to $1/\omega$, since for longer times the phase of the integrand on the right-hand side of this equation varies rapidly with t, so that the contribution to the integral is small. Thus, for instance, in a system with a response time of 10^{-10} sec, $\varepsilon(\omega)$ will show little dispersion at frequencies below 10^{10} Hz. In such a case, for fields varying much more slowly than this the conventional equation $\mathbf{D} = \varepsilon\mathbf{E}$ is a valid approximation to the exact equation.

4.4   The Fluctuation–Dissipation Theorem

The dispersion relations and sum rules that we derived in Sections 4.1 and 4.3 for the generalized susceptibility $Y(\omega)$ are very useful for the examination of experimental data. However, they are not its only important general properties. As we saw in Eq. (90), the imaginary part of $Y(\omega)$ determines the rate of dissipation of energy in a system exposed to an external field. It turns out that this dissipation is related to the quite different problem of the fluctuations of a system in thermodynamic equilibrium. The connection is provided by the fluctuation–dissipation theorem, which we now proceed to derive and discuss. While this theorem can impose restrictions on the possible form of $Y(\omega)$, as we discussed at the end of Section 4.1, its importance is not restricted to this.

Fluctuations of Extensive Variables

We start by considering some general properties of fluctuations. A macroscopic system is usually described in terms of the mean values $\langle x \rangle$ of quantities x whose true value fluctuates about this mean. The deviations of x from $\langle x \rangle$ cannot be characterized by the average value of $x - \langle x \rangle$, since from the definition of the mean this average must vanish if it is taken over a sufficiently large number of observations. The average value of $|x - \langle x \rangle|$, while it differs from zero, is not very convenient to use since it is not a differentiable function of x at $x = \langle x \rangle$. Instead, it is convenient to characterize the fluctuations by their mean square deviation,

$$\Delta = \langle (x - \langle x \rangle)^2 \rangle = \langle x^2 \rangle - \langle x \rangle^2. \qquad (4.140)$$

The ratio of the square root of this quantity to the mean value of x is a dimensionless parameter that determines the relative magnitude of the fluctuations,

$$\delta = \Delta^{1/2}/\langle x \rangle. \qquad (4.141)$$

A large number of physical quantities are extensive variables, i.e., their value for a system is just the sum of their values for each part of the system. For such quantities, it is often convenient to divide the system (for the purposes of analysis) into N macroscopic regions V_k, in each of which the mean value of x is x_0 and its mean square deviation is Δ_0, while the regions are such that the fluctuations in them are statistically independent. For the system as a whole, $\langle x \rangle = N x_0$. Hence, if we denote by x_j the instantaneous value of x in V_j,

$$\Delta = \left\langle \left[\left(\sum_j x_j \right) - N x_0 \right]^2 \right\rangle = \left\langle \left[\sum_j (x_j - x_0) \right]^2 \right\rangle$$

$$= \sum_j \langle (x_j - x_0)^2 \rangle + \sum_{j \neq k} \langle (x_j - x_0)(x_k - x_0) \rangle$$

$$= N \Delta_0, \tag{4.142}$$

since the sum with $j \neq k$ vanishes for statistically independent fluctuations. Thus, the mean square deviation is an extensive variable. On the other hand, the relative magnitude of the fluctuations, $\delta = N^{-1/2} \Delta_0^{1/2} x_0^{-1}$, decreases rapidly as the size of the system increases. It is this fact that allows us to make use, for most purposes, of average values in the description of a system containing a large number of particles. However, in spite of their small relative magnitude, fluctuations can be quite important. For instance, in an isolated system of particles in mechanical equilibrium the mean momentum $\langle \mathbf{p} \rangle$ is zero. However, according to the law of equipartition of energy the average energy of a particle is $\frac{3}{2} k_B T$, and so the mean square momentum of each particle (and hence the mean square deviation of its momentum, since $\langle \mathbf{p} \rangle = 0$) must be proportional to the absolute temperature T.

Fluctuations are of particular importance in the vicinity of order–disorder phase transitions (cf. Sections 2.5 and 3.3). Here, the assumption that fluctuations in different parts of the system are statistically independent is not generally valid. Instead, a correlation appears between fluctuations at points within a characteristic distance of each other. This distance, which is called the correlation length, increases as the phase transition is approached. While the fluctuations are, of course, still much smaller than the mean values of the corresponding quantities, i.e., $\delta \ll 1$, they are much greater than at points far from the phase transition. For instance, in some model systems, δ is proportional to $N^{-1/4}$ in the vicinity of a phase transition, instead of to $N^{-1/2}$.

The inclusion of fluctuations can be regarded as providing an additional description of systems, intermediate between the completely macroscopic one, which involves only mean quantities such as energy, temperature, and pressure, and the full microscopic one in which the coordinates and momentum of each particle are specified. The mean square deviations of the main thermodynamic quantities can usually be expressed in terms of their average thermodynamic values and the derivatives of these. However, we are more interested in the correlation of fluctuations at different times, since it is the time correlation functions that appear in the fluctuation–dissipation theorem.

Time Correlation Functions

A correlation must exist between fluctuations of the same quantity (or of two different, but related quantities) at different times, since it requires a finite time (the relaxation time) for a fluctuation to be produced or to disappear. For classical quantities $A(t)$ and $B(t)$, we can characterize this correlation by the average value of a product,

$$f_{AB}(t, t') = \langle A(t)B(t')\rangle. \tag{4.143}$$

For the sake of convenience, we assume from now on that in thermal equilibrium $\langle A \rangle = \langle B \rangle = 0$, i.e., that A and B represent deviations from average values [even though one cannot then define the relative magnitude of the fluctuations by Eq. (141)]. Since time is isotropic and time-reversal symmetry exists on a microscopic scale (as discussed in Section 3.2), it follows that these correlation functions satisfy the equation

$$f_{AB}(t, t') = f_{AB}(t - t') = f_{AB}(t' - t). \tag{4.144}$$

For quantum-mechanical quantities, A and B must be replaced by the appropriate time-dependent (Heisenberg) operators $\hat{A}(t)$ and $\hat{B}(t)$. Since these do not in general commute, it is natural to define the correlation function as the symmetric function

$$f_{AB}(t - t') = \tfrac{1}{2}\langle \hat{A}(t)\hat{B}(t') + \hat{A}(t')\hat{B}(t)\rangle. \tag{4.145}$$

As an example, let us consider fluctuations in a system such that each small region subject to fluctuations is in internal equilibrium, but the equilibrium state varies from one region to another. In addition, we assume that these fluctuations can be completely described by the values of some quantity x.† Thus, if fluctuations of other thermodynamic quantities take place, their relaxation times are assumed to be much less than that of the quantity x itself; such fluctuations are called thermodynamic or quasi-static fluctuations. If this is the case, then all quantities, including dx/dt, depend only on x. Since $\langle x \rangle = 0$, and for an equilibrium state dx/dt vanishes at the origin, we expand dx/dt as a power series in x, and for small fluctuations retain just the first term, as in Eq. (3.26). Thus, we write

$$dx/dt = -\lambda x, \qquad t > 0. \tag{4.146}$$

If x is a classical quantity, then on solving Eq. (146) we find that the time correlation function, which must be an even function of t, is given by

$$f_{xx}(t) = \langle x(t)x(0)\rangle = \langle x^2 \rangle e^{-\lambda|t|}. \tag{4.147}$$

† Note that in this section, in contrast to Section 4.3, Eqs. (91)–(104), x is not a position coordinate, so that x_{nm} does not have the same meaning in Eq. (155) as it did in Eq. (96).

Hence, the Fourier transform $F_{xx}(\omega)$ of $f_{xx}(t)$, defined as usual by the formula

$$F_{xx}(\omega) = \int_{-\infty}^{\infty} f_{xx}(t)e^{i\omega t}\, dt, \tag{4.148}$$

is given by

$$F_{xx}(\omega) = 2\lambda\langle x^2\rangle/(\omega^2 + \lambda^2). \tag{4.149}$$

On integrating Eq. (149) over frequency, and using Eq. (147) for $t > 0$, we obtain the result

$$f_{xx}(0) = \langle x^2\rangle = \frac{1}{2\pi}\int_{-\infty}^{\infty} F_{xx}(\omega)\, d\omega. \tag{4.150}$$

It is convenient to describe fluctuations by formally introducing into Eq. (146) a hypothetical external force which, if present, would lead to the appearance of these fluctuations. Thus we write

$$dx/dt = -\lambda x + q(t), \tag{4.151}$$

and choose the random force $q(t)$ in such a way as to obtain the correct result for the correlation function $\langle x(t)x(0)\rangle$. For the thermodynamic fluctuations considered in Eq. (147), the correlation function for this random force must be of the form

$$\langle q(t)q(0)\rangle = 2\lambda\langle x^2\rangle\, \delta(t). \tag{4.152}$$

For thermodynamic fluctuations, this random force is fictitious. However, for the interaction of a system of particles with a radiation field that we discuss below, the random force can be attributed to the force on the particle arising from random fluctuations of the electric field in a vacuum.

We note that Onsager's symmetry principle, considered in Section 3.1, is also related to thermodynamic fluctuations, but for systems in which the departure from equilibrium is determined by the set of quantities $x_1, x_2, \ldots, x_n$ rather than by a single quantity x.

The Fluctuation–Dissipation Theorem

In order to obtain a general expression for $f_{xx}(t)$, we must perform a statistical average for a system at finite absolute temperature T. Although this can in principle be done for a classical system, it is more natural to use a quantum-mechanical approach, since the derivation involves microscopic quantities. We consider a canonical ensemble with distribution function

$$\rho(E_n) = \exp((F - E_n)/k_B T). \tag{4.153}$$

where F is the Gibbs free energy and E_n is the nth eigenvalue of the Hamiltonian operator $\hat{H}_0$. The Heisenberg operators, such as those used in Eq. (145), are related to the time-independent Schrödinger operators by the equation

$$\hat{x}(t) = \exp(iH_0 t/\hbar)\hat{x} \exp(-iH_0 t/\hbar). \tag{4.154}$$

If we denote by x_{nm} the matrix element $\langle n|\hat{x}|m\rangle$ of $\hat{x}$ between states n and m, we find from Eqs. (145) and (154) that

$$\bar{f}_{xx}(t) = \sum_{n, m} \tfrac{1}{2}\rho(E_n)(x_{nm}x_{mn}e^{-i\omega_{mn}t} + x_{mn}x_{nm}e^{-i\omega_{nm}t}), \tag{4.155}$$

where an overbar denotes a statistical average at finite temperature, and as usual, $\hbar\omega_{mn} = E_m - E_n$. On interchanging the indices m and n in the second term on the right-hand side of Eq. (155), and performing the Fourier transform according to Eq. (148), we find that

$$\bar{F}_{xx}(\omega) = \pi \sum_{n, m} [\rho(E_n) + \rho(E_m)]|x_{mn}|^2 \delta(\omega - \omega_{mn})$$

$$= \pi \sum_{m, n} \rho(E_n)[1 + \exp(-\hbar\omega_{mn}/k_B T)]|x_{mn}|^2 \delta(\omega - \omega_{mn})$$

$$= \pi[1 + \exp(-\hbar\omega/k_B T)] \sum_{n, m} \rho(E_n)|x_{mn}|^2 \delta(\omega - \omega_{mn}). \tag{4.156}$$

Our next step is to consider the response of the system to an external influence, which we describe by the generalized susceptibility $Y(\omega)$. If $Q(\omega)$ denotes the Fourier component of the external force, we can write, as in Eq. (25),

$$X(\omega) = Y(\omega)Q(\omega). \tag{4.157}$$

Let us assume that this force is a mechanical one, so that its effect can be taken into account by introducing a perturbation $\hat{H}_1$ to the Hamiltonian operator. If we consider a real monochromatic external force, or equivalently if we take account only of the Fourier components of frequency $\pm\omega$ of an arbitrary force acting on the system, we can write, in analogy to Eq. (86),

$$\hat{H}_1 = -\hat{x}(t)(\tfrac{1}{2}q_0 e^{-i\omega t} + \tfrac{1}{2}q_0^* e^{i\omega t}). \tag{4.158}$$

This perturbation induces quantum transitions between the states of the unperturbed Hamiltonian H_0. A quantum of energy $\hbar\omega_{mn}$ is emitted or absorbed in each transition from state n to state m. The transition probability per unit time M_{mn} is determined by Fermi's golden rule, Eq. (93), which in this case has the form

$$M_{mn} = (\pi|q_0|^2/2\hbar^2)|x_{mn}|^2 \delta(\omega \pm \omega_{mn}). \tag{4.159}$$

Hence, the rate of absorption of energy by the system is just

$$W = \frac{\pi}{2\hbar}|q_0|^2\omega \sum_{m,n} |x_{mn}|^2[\delta(\omega + \omega_{mn}) - \delta(\omega - \omega_{mn})]. \quad (4.160)$$

On comparing this equation with Eq. (90), which expressed the rate of absorption of energy in terms of $Y''(\omega)$, we see that

$$Y''(\omega) = \frac{\pi}{\hbar} \sum_{m,n} |x_{mn}|^2[\delta(\omega + \omega_{mn}) - \delta(\omega - \omega_{mn})]. \quad (4.161)$$

Equation (161) differs in form from Eq. (102), because it refers to the net absorption of energy by the system as a result of all transitions, while Eq. (102) described the absorption by an atom in a given state j as a result of transitions that absorb energy.

We now take the statistical average of Eq. (161), remembering that a transition from state m to state n is possible only if state m is initially occupied. By a procedure analogous to that by which Eq. (156) was derived from (153), we find that

$$\bar{Y}''(\omega) = \frac{\pi}{\hbar}\left[1 - \exp\left(-\frac{\hbar\omega}{k_{\mathrm{B}}T}\right)\right] \sum_{m,n} \rho(E_n)|x_{mn}|^2 \delta(\omega - \omega_{mn}). \quad (4.162)$$

Finally, a comparison of Eqs. (156) and (162) shows that

$$\bar{F}_{xx}(\omega) = \hbar\bar{Y}''(\omega)\coth(\hbar\omega/2k_{\mathrm{B}}T). \quad (4.163)$$

We note that Eq. (163) involves only macroscopically observable quantities, as we have used Eqs. (156) and (162) to eliminate the microscopic quantities $|x_{mn}|^2$. On substituting from Eq. (163) in Eq. (150), we find that, since $\bar{Y}''(z)$ and coth z are even functions of z,

$$\langle x^2 \rangle = \frac{\hbar}{\pi} \int_0^\infty \bar{Y}''(\omega)\coth\left(\frac{\hbar\omega}{2k_{\mathrm{B}}T}\right) d\omega. \quad (4.164)$$

Equations (163) and (164) are the fluctuation–dissipation theorem for the quantum case.

In order to pass to the classical limit, we must take the limit of these equations as $\hbar \to 0$. Thus, in Eq. (163) we put $k_{\mathrm{B}}T \gg \hbar\omega$ and so obtain

$$\bar{F}_{xx}(\omega) = (2k_{\mathrm{B}}T/\omega)\bar{Y}''(\omega). \quad (4.165)$$

In Eq. (164), we assume in addition that $\bar{Y}''(\omega) \approx 0$ for $\hbar\omega \gg k_{\mathrm{B}}T$ and so find that

$$\langle x^2 \rangle = \frac{2k_{\mathrm{B}}T}{\pi} \int_0^\infty d\omega\, \frac{\bar{Y}''(\omega)}{\omega}. \quad (4.166)$$

According to the Kramers–Kronig relation of Eq. (31), the integral on the right-hand side of Eq. (166) is related to $\bar{Y}'(0)$, the static value of the real part of $\bar{Y}$. Since $\bar{Y}''(0) = 0$, while $\bar{f}_{xx}(\infty)$ [which corresponds to y_0 of Eq. (31)] also vanishes, it follows from Eqs. (31) and (166) that in the classical case

$$\langle x^2 \rangle = k_B T \bar{Y}(0). \tag{4.167}$$

Application of the Fluctuation–Dissipation Theorem: Energy Density of Radiation Field

In order to use the fluctuation–dissipation theorem in practice, we must first choose the parameters x and q. This can be done from the expression for the rate of absorption of energy, cf. Eqs. (88) and (89), which in our case has the form $dE/dt = -x \, dq/dt$. The relationship between q and x will then determine what is the generalized susceptibility Y whose imaginary part is involved in the fluctuation–dissipation theorem. This procedure is analogous to that used in Section 3.1 in connection with Onsager's principle.

As an example of the application of the theorem, we consider the energy density in an isotropic radiation field which is in equilibrium with some material, regarded as a system of independent charged harmonic oscillators. The equation of motion of such an oscillator subject to an external force $q(t)$ and radiation damping is

$$m \, d^2x/dt^2 + m\omega_0^2 x + (2e^2/3c^3) \, d^3x/dt^3 = q(t). \tag{4.168}$$

Radiation damping is one of the few examples in physics of a force proportional to the third derivative with respect to time of a coordinate. The rate of absorption of energy is given by the product of force and velocity, so that if $q(t)$ and x are both proportional to $\exp(-i\omega t)$, then,

$$dE/dt = -q(t) \, dx/dt = i\omega q(t)x = -x \, dq/dt. \tag{4.169}$$

Hence, our previous theory can be applied, and we write

$$X(\omega) = Y(\omega)Q(\omega). \tag{4.170}$$

On solving Eq. (168) for the case when both q and x are of frequency ω, we readily find that

$$[Y'(\omega) - iY''(\omega)]/|Y(\omega)|^2 \equiv 1/Y(\omega) = m(\omega_0^2 - \omega^2) - 2ie^2\omega^3/3c^3. \tag{4.171}$$

Hence, from the fluctuation–dissipation theorem, Eq. (163), we see that in our case

$$F_{xx}(\omega) = |Y(\omega)|^2 (2\hbar e^2 \omega^3/3c^3) \coth(\hbar\omega/2k_B T). \tag{4.172}$$

This equation can be written in another form, in which we replace the fluctuations of x by those of the random force $q(t)$, or of the random electric field $E(t) = q(t)/e$ that produces the force. Thus, in view of Eq. (170) and with an obvious notation,

$$F_{EE}(\omega) = F_{xx}(\omega)/e^2 \, |Y(\omega)|^2 = (2\hbar\omega^3/3c^3) \coth(\hbar\omega/2k_B T). \quad (4.173)$$

Hence, on integrating this equation over ω just as in the passage from Eq. (163) to (164), we find that

$$\langle E^2 \rangle = \frac{2\hbar}{3\pi c^3} \int_0^\infty \omega^3 \coth\left(\frac{\hbar\omega}{2k_B T}\right) d\omega$$

$$= \frac{4}{3\pi c^3} \int_0^\infty \left[\frac{\hbar\omega}{2} + \frac{\hbar\omega}{\exp(\hbar\omega/k_B T) - 1}\right] \omega^2 \, d\omega. \quad (4.174)$$

The energy density associated with an electric field $\mathbf{E}$ is $U = \langle \mathbf{E} \cdot \mathbf{E} \rangle/4\pi$. Hence, since E in Eqs. (173) and (174) refers to the component of the electric field in the x direction, for an isotropic radiation field

$$U = \frac{3\langle E^2 \rangle}{4\pi} = \frac{1}{\pi^2 c^3} \int_0^\infty \left[\frac{\hbar\omega}{2} + \frac{\hbar\omega}{\exp(\hbar\omega/k_B T) - 1}\right] \omega^2 \, d\omega \quad (4.175)$$

This result is just the well-known Planck formula for the radiation density, including the infinite zero-point contribution. Of course, there are many other ways of deriving this law for the radiation density in equilibrium with some material, but our derivation is a good example of how to apply the fluctuation–dissipation theorem.

*Time Correlation Functions and Transport Coefficients

Further developments of the fluctuation–dissipation theorem, by Kubo and others, have led to some interesting and useful expressions for the transport coefficients in terms of the time-dependent correlation functions. In order to derive these for classical systems, we consider a system which at time $t = -\infty$ was in equilibrium, with a distribution function ρ_0. The system was then exposed to an external force $\mathbf{q}(t)$, whose effect can be taken into account by adding to the Hamiltonian a term

$$\hat{H}_1 = -\hat{A}(\mathbf{r}_1, \mathbf{r}_2, \ldots, \mathbf{r}_N, \mathbf{p}_1, \ldots, \mathbf{p}_N)\mathbf{q}(t), \quad (4.176)$$

where the particles of the system have coordinates $\mathbf{r}_k$ and momenta $\mathbf{p}_k$. As a result of this disturbance, the system's distribution function becomes

$\rho_0 + \delta\rho$. Provided that $\mathbf{q}(t)$ is sufficiently small, we can write

$$\delta\rho = \sum_{k=1}^{N} \left(\delta\mathbf{r}_k(t) \cdot \frac{\partial\rho}{\partial\mathbf{r}_k} + \delta\mathbf{p}_k(t) \cdot \frac{\partial\rho}{\partial\mathbf{p}_k} \right)_0$$

$$= \sum_{k=1}^{N} \int_{-\infty}^{t} d\tau \left(\delta\mathbf{v}_k(\tau) \cdot \frac{\partial\rho}{\partial\mathbf{r}_k} + \mathbf{q}_k(\tau) \cdot \frac{\partial\rho}{\partial\mathbf{p}_k} \right)_0, \qquad (4.177)$$

since $\dot{\mathbf{r}}_k = \mathbf{v}_k$ and, according to Newton's second law, $\dot{\mathbf{p}}_k = \mathbf{q}_k$, the force on particle k. The subscript 0 denotes evaluation for $\rho = \rho_0$. Now, according to Hamilton's equations of motion,

$$\partial H/\partial\mathbf{p}_k = \mathbf{v}_k, \qquad \partial H/\partial\mathbf{r}_k = -\mathbf{q}_k. \qquad (4.178)$$

Hence, in terms of the Poisson bracket

$$[A, B] = \sum_k \left(\frac{\partial A}{\partial\mathbf{p}_k} \frac{\partial B}{\partial\mathbf{r}_k} - \frac{\partial A}{\partial\mathbf{r}_k} \frac{\partial B}{\partial\mathbf{p}_k} \right), \qquad (4.179)$$

Eq. (177) can be written in the simple form

$$\delta\rho = \int_0^\infty [\rho, \hat{H}_1(t - \tau)]_0 \, d\tau. \qquad (4.180)$$

The average value of any microscopic quantity B is given by

$$\bar{B} = \int (\rho_0 + \delta\rho)B \, d\Gamma = \int B \, \delta\rho \, d\Gamma, \qquad (4.181)$$

where $d\Gamma$ is an element of phase space, provided that the mean value of B in equilibrium $\bar{B}_0$ vanishes. Hence, if we substitute in Eq. (181) from Eqs. (176) and (180), we can write

$$\bar{B} = \int_0^\infty \phi_{BA}(\tau)q(t - \tau) \, d\tau, \qquad (4.182)$$

where the linear response function ϕ_{BA} is defined by

$$\phi_{BA}(t) = \int d\Gamma [\rho, A]_0 \, B(t). \qquad (4.183)$$

If the equilibrium distribution function ρ_0 is the canonical one of Eq. (153), then

$$[\rho, A]_0 = \beta\dot{A}\rho_0, \qquad \text{where} \quad \beta = 1/k_B T, \qquad (4.184)$$

and so

$$\phi_{BA}(t) = \beta\overline{\dot{A}(0)B(t)} = -\beta\overline{A(0)\dot{B}(t)}, \qquad (4.185)$$

where in deriving the second equality we used the fact that in equilibrium $(d/dt)[\overline{A(0)B(t)}] = 0$.

Let us consider now a monochromatic external force $q(t) = Q_0 \exp(-i\omega t)$. The response to this force can be written as

$$\bar{B} = \text{Re}\{\chi_{BA}(\omega)Q_0 \exp(-i\omega t)\}, \tag{4.186}$$

where $\chi_{BA}(\omega)$ is the susceptibility that connects B with A. On comparing this expression with Eq. (182), we see that $\chi_{BA}(\omega)$ is just the Fourier transform of the response function,

$$\chi_{BA}(\omega) = \int_0^\infty \phi_{BA}(t) \exp(i\omega t)\, dt. \tag{4.187}$$

The Electrical Conductivity

Although the derivation of Eqs. (182), (185), and (187) is quite complicated, there are several simple examples of their application. For instance, for a system of charged particles in an external electric field, the perturbation to the Hamiltonian has the form

$$H_1 = -\sum_{k=1}^N q_k \mathbf{r}_k \cdot \mathbf{E}(t), \tag{4.188}$$

where the kth particle has charge q_k and position vector $\mathbf{r}_k$, and there are N particles per unit volume. The electrical conductivity connects the average electric current density $\bar{\mathbf{j}} = \overline{\sum_k q_k \dot{\mathbf{r}}_k}$ with the field according to

$$\bar{\mathbf{j}} = \sigma \mathbf{E}. \tag{4.189}$$

A comparison of Eqs. (188) and (189) with (176) and (182), respectively, shows that if $\mathbf{A} = \sum_k q_k \mathbf{r}_k$ and $\mathbf{B} = \sum_k q_k \dot{\mathbf{r}}_k$, then σ corresponds to the response function $\phi_{BA}(t)$, which in our case is a second rank tensor. We now drop the subscripts AB, and instead denote by ϕ_{lm} the lm component of this tensor. From Eq. (185),

$$\phi_{lm}(t) = \beta\langle j_l(0)j_m(t)\rangle. \tag{4.190}$$

Hence, on substituting from Eq. (190) in Eq. (187), we obtain for the electrical conductivity the formula

$$\sigma_{lm}(\omega) = \beta \int_0^\infty dt\, \exp(i\omega t)\langle j_l(0)j_m(t)\rangle. \tag{4.191}$$

Thus, the electrical conductivity is determined by the Fourier transform of the current–current time correlation function.

The simplest case to which Eq. (191) can be applied is that of a system of N classical particles per unit volume, each of charge e, moving independently of each other. In that case, we find from Eq. (191) that, since the velocity–velocity time correlation function $\langle v_l(0)v_m(t)\rangle$ is the same for each particle,

$$\sigma_{lm}(0) = Ne^2\beta \int_0^\infty \langle v_l(0)v_m(t)\rangle \, dt. \tag{4.192}$$

Since, according to Eq. (47), the integral in Eq. (192) is just the diffusion coefficient D_{lm}, the above result is the well-known Einstein relation

$$\sigma_{lm}(0) = Ne^2 D_{lm}/k_B T. \tag{4.193}$$

We note that, for a system of interacting particles, the Einstein relation has the more general form

$$\sigma_{lm}(0) = e^2 D_{lm} \overline{\Delta N^2}/k_B T, \tag{4.194}$$

where $\overline{\Delta N^2}$ is the spatial average of the square fluctuation in the number density of particles. Equation (194) reduces to the Einstein relation only for classical noninteracting particles, for which $\overline{\Delta N^2} = N$.

The Electrical Susceptibility of a Dielectric Medium

Another simple example of the application of response functions to the calculation of macroscopic properties is provided by the electrical susceptibility of a dielectric medium. We consider a medium (solid or liquid) containing molecular permanent electric dipoles $\mathbf{p}_k$ that can rotate (or change direction in some other manner) under the influence of an external electric field. For this system,

$$H_1 = -\sum_{k=1}^{N} \mathbf{p}_k \cdot \mathbf{E}(t), \tag{4.195}$$

while the electrical susceptibility χ_e is defined by

$$\mathbf{P} = \sum_{k=1}^{N} \mathbf{p}_k = \chi_e \mathbf{E}, \tag{4.196}$$

where there are N such dipoles per unit volume. It follows that in this case $A = B = \Sigma_k \mathbf{p}_k$. Hence, in view of Eqs. (185) and (187), the lm component of the tensor susceptibility is given by

$$\chi_e(\omega)_{lm} = -\beta \int_0^\infty dt \, \exp(i\omega t) \sum_j \sum_k \overline{p_j(0)_l p_k(t)_m}. \tag{4.197}$$

where $p_j(0)_l$ denotes the component of $\mathbf{p}_j$ in direction l at time $t = 0$.

An approximation often used to evaluate the correlation function appearing in the integrand of the above equation is to assume that the dipole moment of any given molecule in a given direction decays exponentially with time, as a result of the interaction with the other molecules, but that each molecule rotates independently of the others. Thus, it is assumed that

$$\dot{p}_k(t)_m = -p_k(t)_m/\tau, \tag{4.198}$$

and

$$\overline{p_j(0)_l p_k(t)_m} = 0 \quad \text{unless} \quad l = m, \quad j = k. \tag{4.199}$$

In that case, it follows from Eq. (197) that χ_e is a scalar, and that

$$\chi_e(\omega) = \chi_e(\omega)_{mm} = \frac{\beta}{\tau} \int_0^\infty dt \, \exp(i\omega t) \sum_k p_k(0)_m^2 \exp(-t/\tau) = \frac{C}{1 - i\omega\tau}, \tag{4.200}$$

where C and τ are in general functions of temperature but not of frequency. Equation (200) is just the Debye formula for the electrical susceptibility, and is a good approximation in many systems. When it is not valid, detailed calculations must be used to evaluate the dipole–dipole time correlation function that appears in Eq. (197).

To conclude this section, we note that this Kubo formalism can readily be extended to quantum-mechanical systems. It has also been generalized to treat thermal disturbances which, unlike mechanical ones, cannot be represented by an additional term H_1 in the Hamiltonian.

Problems

4.1 Check that the S-matrix given by Eq. (60) satisfies the unitarity condition of Eq. (10).

4.2 Derive the S-matrix for a one-dimensional system in which the potential energy $V(x) = \lambda\delta(x)$, and check that the result satisfies the unitarity conditions of Eq. (10). See I. R. Lapidus, *Amer. J. Phys.* **37**, 1064 (1969). See also Problem 5.6.

4.3 For the Coulomb potential $V(r) = Ze^2/r$, the asymptotic form of the radial part of the solution of Schrödinger's equation as $r \to \infty$ has the form

$$\psi_l(r) \sim \sin(kr - l\pi/2 - \zeta \ln 2kr + \eta_l)/r,$$

where

$$\zeta = Zme^2/\hbar^2 k \quad \text{and} \quad \exp(2i\eta_l) = \Gamma(l + 1 + i\zeta)/\Gamma(l + 1 - i\zeta).$$

(a) Find the scattering amplitude $f(k)$ and scattering matrix $S(k)$ for this system.

(b)   Show that the scattering cross section $d\sigma = |f(\theta)|^2 \, d\Omega$ has the well-known Rutherford form

$$d\sigma = (Ze^2 m/2\hbar^2 k^2)^2 \, d\Omega/\sin^4(\theta/2).$$

(c)   By writing down $S(k)$ for complex k, show that the bound states have energies

$$E_n = -m(Ze^2)^2/(4\hbar^2 n^2), \qquad n = 1, 2, \ldots .$$

See Ref. [35, Section 136].

4.4   Show by direct integration that the dielectric constant $\varepsilon(\omega)$ given by Eq. (18) satisfies the Kramers–Kronig relations, Eqs. (34) and (35), and the sum rule of Eq. (36).

4.5   It is often convenient to write the relationship between the electric displacement **D** and the electric field **E** for an isotropic insulator in the form

$$\mathbf{D}(t) = \varepsilon_\infty \mathbf{E}(t) + (\varepsilon_s - \varepsilon_\infty) \int_{-\infty}^{t} \mathbf{E}(u)\phi(t - u) \, du,$$

where ε_s and ε_∞ denote, respectively, the static ($\omega = 0$) and high-frequency ($\omega \to \infty$) values of the dielectric constant. [Incidentally, if $\phi(u)$ represents the delayed response due to the motion of atoms or molecules, ε_∞ will include the effects of polarization due to electrons, and so will not equal 1 in Gaussian units or ε_0 in S.I. units.]

(a)   Find an expression for $\varepsilon(\omega)$, and show that $\phi(u)$ is a normalized response function, i.e., $\int_0^\infty \phi(u) \, du = 1$.

(b)   Derive an expression for the polarization current density, $j_0(t) = (1/A) \, \partial \mathbf{D}/\partial t$, at time t after the application of an electric field of unit strength, and hence show that

$$\varepsilon''(\omega) = \frac{A}{\omega} \int_{-\infty}^{\infty} j_0\!\left(\frac{u}{\omega}\right) \sin u \, du,$$

where $A = 1$ in S.I. units and 4π in Gaussian units.

(c)   It is often found that, for a wide range of times, $j_0(t) = Bt^{-m}$, $0 < m < 1$. Show that if this relation were true for all times, then

$$\varepsilon''(\omega) = (AB/\omega)j_0(a/\omega), \qquad \text{where} \quad a = [\Gamma(1 - m)\cos(m\pi/2)]^{-1/m}.$$

For $0.3 < m < 1$, show that $a = 0.63 \pm 0.02$. This result is known as Hamon's approximation, and is of great value in the analysis of experimental results.

(d)   Show that the results of (c) imply that $\varepsilon''(\omega)$ is proportional to $\omega^{(m-1)}$, $0 < m < 1$, and hence, from tables of Hilbert transforms, that

$$|\varepsilon''(\omega)/[\varepsilon'(\omega) - \varepsilon_\infty]| = \cot(m\pi/2),$$

a result that is often observed experimentally over limited frequency ranges. See G. Williams, *Trans. Faraday Soc.* **58**, 1041 (1962); A. K. Jonscher, *Nature* **267**, 673 (1977).

4.6 (a) Show that, because of analytic considerations, the form for the dielectric constant

$$\varepsilon(\omega) = \varepsilon_\infty + G\omega^{m-1}$$

obtained in Problem 4.5 cannot be valid in the limits $\omega \to 0$ or $\omega \to \infty$.
(b) Assuming that the error in the above expression at low frequencies has a negligible effect on the integrals involved in the sum rules, derive from Eq. (115) the maximum frequency ω_0 up to which this expression can be valid. Note that, from Eq. (90), $\varepsilon''(\omega)$ is nonnegative.
(c) Derive additional sum rules for $\varepsilon(\omega) - \varepsilon_\infty$ from the relationships obtained in Eqs. (130) and (131), and investigate whether these impose additional restrictions on the maximum frequency ω_0.

*4.7 A simple example of the connection between fluctuations and energy dissipation is that of a one-dimensional damped harmonic oscillator subject to a random fluctuating force $q(t)$. We write its equation of motion in the form $m\ddot{x} + a\dot{x} + Kx = q(t)$.
(a) Take the Fourier transform of this equation to obtain $X(\omega)$, and hence obtain $\langle x^2 \rangle$ as a function of $|Q(\omega)|^2$.
(b) If $|Q(\omega)|^2$ is a constant, or if it is sharply peaked at a resonant frequency ω_R, it can be taken outside the integral involved in $\langle x^2 \rangle$. Show that in such a case

$$\langle x^2 \rangle = |Q(\omega_R)|^2/4aK.$$

(c) If the oscillator is in thermal equilibrium with a reservoir at temperature T, then its mean potential energy $\frac{1}{2}K\langle x^2 \rangle = \frac{1}{2}k_B T$. By equating the value of $\langle x^2 \rangle$ obtained from this relationship with that obtained in (b), show that $|Q(\omega_R)|^2$ is proportional to $ak_B T$. Since ω_R can be chosen arbitrarily, this implies that a damping force $a\dot{x}$ is associated with a fluctuating force such that the energy density in any frequency range $\Delta\omega$ is proportional to $ak_B T \, \Delta\omega$. This random force is just that required to provide the energy dissipated by the damping. For an electrical circuit, this result corresponds to Nyquist's theorem. See C. W. McCombie, *Rep. Progr. Phys.* **16**, 266 (1953).

Chapter 5 | The Method of the Small Parameter

5.1 Introduction

So far, we have discussed only the qualitative form of the solution of physical problems. However, in order to complete a qualitative analysis of the problem, we also need to obtain an idea of the order of magnitude of the solution. This can be a matter of great practical importance. For instance, if a person is told than an animal belonging to the cat family, with whiskers on its face, stripes on its back, and retractable claws in its padded feet, is coming round the corner, he may still want to know whether the animal is a small striped domestic cat or a large tiger. Similarly, it can be important to know whether a small perturbation to a system will produce only a small effect or a quite drastic one. For instance, a small alteration in the shape or temperature of the core of a nuclear reactor may change slightly the average number of neutrons produced by each neutron released in a reaction, and the sign and magnitude of this change will determine whether the reactor will tend to stop working or to explode as a result. In this chapter, we describe some methods of deriving the magnitude of the solutions (i.e., approximate numerical values) for problems that usually cannot be solved exactly in closed form.

A Typical Problem

A simple example of such a problem, various cases of which will be considered in this section, is that described by the equation

$$m\, d^2y/dt^2 + a\, dy/dt + (K + bf(t))y = g(t), \qquad (5.1)$$

where a and b are small and $f(t)$ is a given function of t, of order of magnitude unity. If t denotes time, this equation describes the motion of a linear damped harmonic oscillator with a time-dependent restoring force plus an external force $g(t)$. We have met various forms of this problem in most of the previous chapters. If, on the other hand, we replace t by a spatial variable x, Eq. (1) describes the vibrations of a string subject to various forces, where the term involving time in the wave equation is replaced by one involving $-\omega^2 y$ and incorporated in Ky. Schrödinger's equation for the wave function of a nearly free particle in one dimension is also of this form, with x replacing t and $a = 0, g = 0$.

The solution of a problem is determined not only by its governing equation, such as Eq. (1), but also by the boundary conditions, and these can crucially affect the form of the solution. A linear second-order ordinary differential equation, such as (1), generally requires two boundary conditions to determine its solution uniquely. However, there is a great difference whether these equations relate to the same point, e.g., with $\dot{y}$ denoting dy/dt,

$$y(0) = p, \qquad \dot{y}(0) = q, \qquad (5.2)$$

or whether they relate to different points, e.g.,

$$y(0) = y(l) = 0. \qquad (5.3)$$

In the first case, nontrivial solutions exist for all values of K. In the second one, if $g(t) = 0$, they only exist for specific values of K, known as the eigenvalues of the problem, while if $g \neq 0$, it can happen that no solution at all exists for these values of K.

The methods that we consider involve expanding the solution as a power series in a parameter. Frequently, this parameter is known on physical grounds to be small, in which case the method is known as the method of the small parameter, but this is not essential. In Section 5.2, we will consider various types of perturbation theory, which in principle do not require the existence of such a small parameter. However, in practice it is most convenient to employ a parameter that for physical reasons must be small, and in Section 5.3 we discuss the choice of such a parameter for various problems. Even when a small parameter exists, though, it may not be simple to use it if it appears as a factor multiplying a large quantity or multiplying the highest derivative of a differential equation. The solution of such problems is considered in Section 5.4. We also consider in this section some problems in which perturbation theory can be applied even though only the form of the zeroth-order solution is known, and not its explicit value.

Perturbation Theory — The Series Expansion Technique

The basic problem that we consider can be expressed formally as that of solving an equation of the form

$$J(y) = g, \tag{5.4}$$

where J is some operator.† We split $J(y)$ into two parts, and write

$$J(y) \equiv L(y) - N(y) = g \tag{5.5a}$$

or equivalently

$$J(y) \equiv L(y) - \lambda N(y) = g. \tag{5.5b}$$

It is usually convenient to choose $L(y)$ to be a linear operator, even if J is itself nonlinear, and we henceforth assume that this is the case. Let y_0 be the unique solution of the equation $L(y) = g$ together with its boundary conditions. We then consider the equation

$$L(y) = g + \lambda N(y), \tag{5.6}$$

and express its solution as a power series in λ. Equation (6) is identical to Eq. (5b), so that if the problem contains a small parameter we express the solution as a power series in it. Otherwise, we introduce λ as a mathematical convenience, and set $\lambda = 1$ at the end to obtain the solution of Eq. (5a). The term $\lambda N(y)$ can be regarded as a perturbation to the operator $L(y)$, and so the method of solution that we now discuss is known as perturbation theory.

In order to solve Eq. (6), we consider first solutions for λ close to zero, and look for a solution of the form

$$y = y_0 + \lambda y_1 + \lambda^2 y_2 + \cdots = \sum_{k=0}^{\infty} \lambda^k y_k. \tag{5.7}$$

Since L is a linear operator, if the infinite series is sufficiently well convergent,

$$L(y) = \sum_{n=0}^{\infty} \lambda^k L(y_k). \tag{5.8}$$

We require that $N(y)$ be analytic in y, so that we can write

$$N(y_0 + \lambda y_1 + \lambda^2 y_2 + \cdots) = N(y_0) + \lambda N_1(y_0, y_1) + \lambda^2 N_2(y_0, y_1, y_2) + \cdots, \tag{5.9}$$

† For simplicity, in this chapter we omit the symbol above operators.

where the coefficient of λ^k depends only on $y_0, y_1, \ldots, y_k$. Then, on substituting from Eq. (9) and (8) into Eq. (6) and equating the coefficients of corresponding powers of λ, we find that

$$L(y_0) = g, \tag{5.10a}$$

$$L(y_1) = N(y_0), \tag{5.10b}$$

$$L(y_2) = N_1(y_0, y_1), \tag{5.10c}$$

$$\vdots$$

$$L(y_k) = N_{k-1}(y_0, y_1, y_2, \ldots, y_{k-1}). \tag{5.10d}$$

This set of equations can, at least in principle, be solved recursively. If the power series of Eq. (7), with the functions y_k determined by Eqs. (10), converges, it then provides a solution to our original problem. Since L is generally chosen to be a simple linear operator, it is usually much easier to solve Eqs. (10) than the original equation.

Solution for Problem with Two Boundary Conditions at the Same Point

A simple example of the application of this series expansion technique is provided by the problem described by Eqs. (1) and (2), with $g(t) = 0$. We choose

$$L(y) = m\, d^2y/dt^2 + Ky, \qquad N(y) = -a\, dy/dt - bf(t)y. \tag{5.11}$$

The solution of Eq. (10a) with the boundary conditions of Eq. (2) is then

$$y_0 = p \cos \omega_0 t + (q/\omega_0) \sin \omega_0 t, \qquad \text{where} \quad \omega_0^2 = K/m. \tag{5.12}$$

For Eqs. (10b)–(10d), the appropriate boundary conditions are just

$$y_k(0) = \dot{y}_k(0) = 0, \qquad k = 1, 2, \ldots. \tag{5.13}$$

On substituting the explicit forms of $y_0, y_1, \ldots, y_{k-1}$ in Eq. (10d) for y_k, we obtain an equation of the form

$$L(y_k) = \phi_k(t). \tag{5.14}$$

Its solution, subject to these boundary conditions, is readily found to be

$$y_k(t) = \frac{1}{m\omega_0} \int_0^t \phi_k(s) \sin[\omega_0(t - s)]\, ds. \tag{5.15}$$

Thus, Eqs. (12) and (15) contain the complete solution of the problem, provided that the resulting infinite series for y converges. This is always the case if the problem is linear in y and $f(t)$ is bounded, as we show in Section 5.2.

Even if the series does converge, it does not necessarily provide a useful solution to the problem. For the series solution to be useful, we require that either the first two or three terms of it provide an adequate approximation to the full solution or that the series can readily be summed to all orders. As a simple example, let us consider the free oscillations of a damped harmonic oscillator starting from rest, so that $b = 0$, $g(t) = 0$, and $q = 0$ in Eq. (1) and (2). We then readily find that, if we put $\lambda = a/m$, the damping parameter per unit mass,

$$y_0 = p \cos \omega_0 t, \qquad y_1 = (p/2\omega_0) \sin \omega_0 t - (p/2)t \cos \omega_0 t, \qquad (5.16)$$

so that the amplitude of y_1 increases without bound as t increases. Since this cannot be true of the exact solution of the problem, we conclude that the approximation $y = y_0 + \lambda y_1$ cannot be valid for all values of t. In fact, we cannot expect such an approximation to be valid unless λy_1 is much less than y_0, i.e., unless $t \ll 2m/a$. If this condition is fulfilled, $y_0 + \lambda y_1$ is, of course, just the first two terms of the expansion in powers of a/m of the exact solution of this problem,

$$y = p \exp(-at/2m)[\cos \omega_e t + (a/2m\omega_e) \sin \omega_e t], \qquad \omega_e = \sqrt{\omega_0^2 - a^2/4m^2}.$$

$$(5.17)$$

Renormalization Techniques

Our analysis so far has been based on the expansion of y as a power series in λ where the individual functions are independent of λ. However, it is sometimes possible to extend the validity of an approximation such as $y = y_0 + \lambda y_1$, and hence the usefulness of perturbation theory, by allowing the functions y_0, y_1, etc., themselves to depend on λ. One way of doing this is to change the independent variable, from t to s, where

$$t = s(1 + c_1 \lambda + c_2 \lambda^2 + \cdots) \qquad (5.18)$$

and the coefficients $c_1, c_2, \ldots$ are suitably chosen. Such a technique, which is known as renormalization of the independent variable, is of value in a wide range of problems, although it can sometimes create difficulties in satisfying the boundary conditions. Another possibility is to express some parameter appearing in y_0, for instance its frequency ω, as a power series in λ,

$$\omega = \omega_0 + \lambda \omega_1 + \lambda^2 \omega_2 + \cdots. \qquad (5.19)$$

This renormalization of the frequency is equivalent to the renormalization of t and is often simpler to use.

As an example of the application of this technique, we consider the motion of an undamped oscillator, starting from rest at time $t = 0$, in a potential field

$$V(y) = V_0 + (K/2)y^2 + (c/4)y^4 \tag{5.20}$$

that is symmetric in y. The equation of motion for this system is

$$m\, d^2y/dt^2 + Ky + cy^3 = 0, \qquad y(0) = p, \qquad \dot{y}(0) = 0, \tag{5.21}$$

which is a generalization of Eq. (1) to a nonlinear problem, and we set our expansion parameter $\lambda = c/m$.

Straighforward application of the techniques of Eqs. (10)–(15) leads to

$$y_0 = p \cos \omega_0 t,$$
$$y_1 = (p^3/32\omega_0^2)(\cos 3\omega_0 t - \cos \omega_0 t) - (3p^3/8\omega_0)t \sin \omega_0 t, \tag{5.22}$$

and the last term in y_1 presents the same problems as those encountered with the damped harmonic oscillator. In order to overcome them, we write $y_0 = p \cos \omega t$, where the renormalized frequency ω is given by Eq. (19), and attempt to choose ω_1 in such a way that y_1 no longer contains a term whose amplitude is proportional to t. We can no longer use Eqs. (10) directly, but can still equate the coefficients of corresponding powers of λ on the two sides of Eq. (6). Thus, in our case we write

$$(d^2/dt^2 + \omega_0^2)(p \cos \omega t + cy_1/m + \cdots) = -(c/m)(p \cos \omega t + cy_1/m + \cdots)^3. \tag{5.23}$$

On substituting for ω from Eq. (19) and equating the coefficients of the first power of c/m, we find that

$$\ddot{y}_1 + \omega_0^2 y_1 = -p^3 \cos^3 \omega_0 t + 2p\omega_0 \omega_1 \cos \omega_0 t$$
$$= -(p^3/4) \cos 3\omega_0 t + (2p\omega_0 \omega_1 - 3p^3/4) \cos \omega_0 t. \tag{5.24}$$

In order that y_1 should not contain a resonance term, i.e., one whose amplitude is unbounded, the coefficient of $\cos \omega_0 t$ on the right-hand side of Eq. (24) must be zero. Thus, we choose

$$\omega_1 = 3p^2/8\omega_0, \tag{5.25}$$

in which case the first two terms in the solution of Eq. (21) are

$$y_0 = p \cos(\omega_0 t + c\omega_1 t/m), \qquad y_1 = (p^3/32\omega_0^2)[\cos 3\omega_0 t - \cos \omega_0 t]. \tag{5.26}$$

As a result of this change of frequency, cy_1/m will be much less than y_0 for all values of t, if c/m is sufficiently small, so that $y = y_0 + cy_1/m$ should be a good approximation to y for all times. This solution is periodic and bounded, as we expect on physical grounds for an undamped oscillator with no external driving force. The approximate solutions, $y = y_0 + cy_1/m$, of Eq. (21), where y_0 and y_1 are given either by Eq. (22) or by (26), are equivalent for small ct/m, since the solution of Eq. (22) corresponded to the expansion of $p \cos(\omega_0 t + c\omega_1 t/m)$ as a power series in ct/m for small ct/m.

$$\cos(\omega_0 t + c\omega_1 t/m) \approx \cos \omega_0 t - (c/m)\omega_1 t \sin \omega_0 t. \qquad (5.27)$$

Eigenvalue Problems

We now turn to problems of the type described by Eqs. (1) and (3), in which the solution is required to satisfy a single boundary condition at each of two points. We restrict our attention to a linear problem without a driving force $g(t)$, in which these conditions are homogeneous, i.e., of the form of Eq. (3) or, more generally, $a_1 y + a_2 \dot{y} = 0$ at each of these points, or in which a periodic solution is required, $y(T) = y(0)$ and $\dot{y}(T) = \dot{y}(0)$. A trivial solution of problems of this sort is $y = 0$, but there may be some values of K, the eigenvalues of the problem, for which a nontrivial solution, the corresponding eigenfunction, exists. Just as in the previous examples we examined how a perturbation term $\lambda N(y)$ affected the known solution y_0 of the unperturbed problem $L(y) = 0$, so we now study how such a perturbation affects the eigenvalues and eigenfunctions of our unperturbed problem. This sort of question arises very frequently in physics, and especially in quantum mechanics.

Our operator $J(y)$ now involves both a known operator H and its eigenvalue K, which we now denote by $-E$ to accord with the usual notation for Schrödinger's equation. Thus, our equation is now

$$J(y) \equiv (H - E)y = 0, \qquad (5.28)$$

and we split the operator H into two parts,

$$H = H_0 + \lambda H_1. \qquad (5.29)$$

Initially, we write in Eq. (5)

$$L(y) = (H_0 - E_0)y, \qquad N(y) = -H_1 y + (E - E_0)y/\lambda, \qquad (5.30)$$

so that Eq. (6) becomes

$$(H_0 - E_0)y = -\lambda H_1 y + (E - E_0)y. \qquad (5.31)$$

The major difference between this problem and those considered previously is that the perturbation term $\lambda N(y)$ is not known in advance, since it contains the perturbed eigenvalue E. The value of E_0 is chosen so that the solution

y_0 of $L(y) = 0$ is known, i.e., E_0 is an eigenvalue of H_0 and y_0 the corresponding eigenfunction (which we assume, for convenience, is not degenerate).

In order to solve Eq. (31), we assume that both y and E can be expanded as power series in λ, so that in addition to Eq. (7) we write

$$E = \varepsilon_0 + \lambda\varepsilon_1 + \lambda^2\varepsilon_2 + \cdots = \sum_{j=0}^{\infty} \lambda^j\varepsilon_j, \tag{5.32}$$

where $\varepsilon_0 = E_0$. This use of two power series to solve Eq. (31) is similar to the use of two such series in the renormalization technique. However, in that case we required a special method to find the coefficients in the two power series, so as to eliminate the divergent terms in the solution, while we will now derive explicit general formulas for the coefficients in each series. Since we have assumed that H is a linear operator, on substituting these expansions in Eq. (31) and equating the coefficients of λ^j on the two sides, we find that

$$(H_0 - E_0)y_j = -H_1 y_{j-1} + \sum_{k=0}^{j-1} \varepsilon_{j-k} y_k, \qquad j > 0. \tag{5.33}$$

Equation (33) is similar to Eq. (10d), and we should like to solve the set of equations for different j iteratively, just as we did with Eqs. (10). However, Eq. (33) involves two unknowns at each stage, y_j and ε_j, and we must now consider how to separate them.

Rayleigh–Schrödinger Perturbation Theory

One technique that is widely used is to express each term y_j of the expansion as a linear combination of the eigenfunctions u_n of H_0, which are such that $u_0 \equiv y_0$ and in general

$$H_0 u_n = E_n u_n. \tag{5.34}$$

We assume that these eigenfunctions are chosen in such a way that they form an orthonormal set over the interval Ω, i.e., so that

$$(u_m | u_n) \equiv \int_\Omega u_m^* u_n \, d\Omega = \delta_{m,n}. \tag{5.35}$$

We denote the matrix element of H_1 between eigenfunctions by H_{mn}^1; hence, in view of Eqs. (29), (34), and (35),

$$(u_m | H | u_n) = E_n \delta_{m,n} + \lambda H_{mn}^1. \tag{5.36}$$

On writing

$$y_j = \sum_n c_n^j u_n \tag{5.37}$$

and on substituting this in Eq. (33) we find that

$$\sum_n c_n^j(E_n - E_0)u_n = -\sum_n c_n^{j-1}H_1u_n + \sum_{k=0}^{j-1} \varepsilon_{j-k} \sum_n c_n^k u_n. \qquad (5.38)$$

We now multiply this equation by u_m^*, integrate over Ω, and substitute from Eqs. (34) and (35) to obtain the equation

$$c_m^j(E_m - E_0) = -\sum_n c_n^{j-1}H_{mn}^1 + \varepsilon_j c_m^0 + \sum_{k=1}^{j-1} \varepsilon_{j-k} c_m^k. \qquad (5.39)$$

For $m = 0$, since $y_0 = u_0$ so that $c_m^0 = \delta_{m,0}$, we find from Eq. (39) that

$$\varepsilon_j = \sum_n c_n^{j-1}H_{on}^1 - \sum_{k=1}^{j-1} \varepsilon_{j-k} c_0^k,$$

which determines ε_j uniquely in terms of the solutions for lower orders, $k < j$. Equation (39) for $m \neq 0$ can then be used to determine all the c_m^j except c_0^j. At each stage, c_0^j is arbitrary, and it is often convenient to require it to be zero. In that case, of course, the second term on the right-hand side of Eq. (40) vanishes.

While the above technique, known as Rayleigh–Schrödinger perturbation theory, can in principle be used to all orders, the formulas become so complicated that in practice it is usual to stop at terms of order λ^2, i.e., at $j = 2$, unless some means can be found of summing the whole series or a subsection of it.

Mathieu's Equation

A simple example of the application of Rayleigh–Schrödinger perturbation theory is provided by Mathieu's equation, which can be written in the form

$$-d^2y/dx^2 + (\lambda \cos x - E)y = 0. \qquad (5.41)$$

This is another special case of Eq. (1), with x replacing t. It can be regarded as Schrödinger's equation for a particle moving in a periodic potential $V(x) = (2m\lambda/\hbar^2) \cos x$, which is one of the simplest possible model potentials for an electron in a one-dimensional crystal. We look for solutions of Eq. (41) that have the same periodicity as $V(x)$, i.e., with $y(2\pi) = y(0)$, $y'(2\pi) = y'(0)$, where y' denotes dy/dx.

In terms of the notation of Eqs. (28)–(31), $H_0 = -d^2/dx^2$ and $H_1 = \cos x$. Before starting to find the eigenvalues and eigenfunctions of H, we note that H is a symmetric function of x, $H(-x) = H(x)$. Hence, in accordance with the discussion in Sections 3.1 and 3.3, the eigenfunctions can be classified as either even or odd functions of x, and these two types can be considered separately. Let us examine only the odd eigenfunctions of Eq. (41). For the eigenfunctions of the unperturbed problem, we write

$$u_n = \pi^{-1/2} \sin(n+1)x, \qquad E_k = (n+1)^2, \qquad n = 0, 1, 2, \ldots, \qquad (5.42)$$

and a simple calculation shows that in this case

$$H^1_{nm} = \tfrac{1}{2}(\delta_{n,m+1} + \delta_{n,m-1}). \qquad (5.43)$$

We note that expanding y as a series in the functions u_k is in this case equivalent to expanding it in a Fourier sine series, as is natural for an odd periodic function. On applying in turn Eqs. (40) and (39), we then readily find that, for the smallest eigenvalue and its eigenfunction,

$$\varepsilon_1 = 0, \qquad c^1_m = -\tfrac{1}{6}\delta_{m,1}, \qquad \varepsilon_2 = -\tfrac{1}{12}, \qquad \text{etc.} \qquad (5.44)$$

so that

$$E = 1 - \lambda^2/12 + \cdots, \qquad y = \pi^{-1/2}(\sin x + (\lambda/6)\sin 2x + \cdots). \qquad (5.45)$$

The derivation of additional terms in E and y is tedious but not difficult.

Brillouin–Wigner Perturbation Theory

An alternative way of expressing the operator $J(y)$ of Eq. (28) in the form $L - \lambda N$ is to choose

$$L(y) = (H_0 - E)y, \qquad N(y) = -H_1 y. \qquad (5.46)$$

Such a choice would seem more natural than that of Eq. (30), since $N(y)$ now only contains the term of $J(y)$ that is explicitly multiplied by λ. However, it has the disadvantage that E is not known in advance, and that no solution exists of $L(y) = 0$ which satisfies the boundary conditions, since an eigenvalue of H will not normally be one of H_0. In order to use Eq. (46), we again choose as our first approximation to y the eigenfunction y_0 with eigenvalue E_0, and write $(H - E)y = 0$ in the form

$$(H_0 - E)(y - y_0) = -\lambda H_1 y + (E - E_0)y_0. \qquad (5.47)$$

The differences between this equation and Eq. (31) [from the left-hand side of which we can subtract $(H_0 - E_0)y_0$ as this equals zero] are that in Eq. (47) the unknown E replaces the known E_0 on the left-hand side, while the known y_0 replaces the unknown y in the second term on the right-hand side. The use of E in place of E_0 means that we are using at each stage a renormalized energy, so that, in contrast to Rayleigh–Schrödinger perturbation theory, at each order of perturbation theory we now have to adjust all the lower-order terms. If we now express y throughout Eq. (47) and E on its right-hand side by the power series of Eqs. (7) and (32), and equate coefficients of corresponding powers of λ, we find that

$$(H_0 - E)y_j = -H_1 y_{j-1} + \varepsilon_j y_0. \tag{5.48}$$

The value of E on the left-hand side of this equation has to be determined self-consistently at each stage. The resulting power series for y is known as the Brillouin–Wigner perturbation series.

If we again express y_j as a linear combination of the eigenfunctions u_n of H_0, as in Eq. (37), and apply the procedure used to derive Eq. (39) from (38), we find now that

$$c_m^j(E_m - E) = -\sum_n c_n^{j-1} H_{mn}^1 + \varepsilon_j \delta_{m,0}. \tag{5.49}$$

Since the normalization of y is not determined in advance, we can choose it so that $(y - y_0)$ is orthogonal to y_0, i.e., so that $c_0^j = \delta_{j,0}$, as previously. In that case, Eq. (49) for $m = 0$ leads to

$$\varepsilon_j = \sum_n c_n^{j-1} H_{0n}^1, \tag{5.50}$$

which is of the same form as Eq. (40) with $c_0^k = 0$ for $k > 0$. However, the coefficients c_m^j are now given by

$$c_m^j = -\sum_n c_n^{j-1} H_{mn}^1 / (E_m - E), \qquad m \neq 0. \tag{5.51}$$

A simple application of Brillouin–Wigner perturbation theory is to a degenerate case where H_0 has an eigenvalue E_0 with two eigenfunctions, u_0 and u_1, while only the matrix elements of H_1 between these two states are non-zero. In that case, $\varepsilon_1 = 0, c_1^1 = -H_{01}^1 / (E_0 - E)$, and so to second order,

$$\lambda^2 \varepsilon_2 = (E - E_0) = -\lambda^2 (H_{01}^1)^2 / (E_0 - E). \tag{5.52}$$

Thus, the renormalized energy

$$E = E_0 \pm \lambda H_{01}^1, \tag{5.53}$$

a result that can only be derived by the Rayleigh–Schrödinger method if the much more complicated form of that perturbation theory for degenerate states is used.

Further developments of perturbation theory, based on the formulation of our problems in terms of integral equations, are considered in Section 5.2. For problems with boundary conditions at two points, such a formulation involves the concept of Green's functions, and some of the properties and uses of these are established and discussed. Another example of the application of perturbation theory is presented at the end of Section 5.3. In this, the zeroth-order operator is not fixed in advance, but instead the division of operator into zeroth order plus perturbation is varied as the solution of the problem proceeds.

Choice of the Small Parameter

The next problem that we consider is that of how to choose the term that is to be treated as a perturbation. In particular, we are interested in finding, where possible, a small parameter of a physical system which can be used as the expansion parameter in order to ensure rapid convergence of the series solution. While in many cases the choice of such a parameter is obvious, in other cases it can be a far from trivial problem. A good example of the latter situation is provided by systems containing large numbers of particles, i.e., many-body problems.

Many-body problems arise not only in physics, but in practically all important scientific and social problems. Unfortunately, even in the physical sciences an exact solution can be found only for the two-body problem, and for systems of three bodies it is already necessary to use some approximations. For this reason, modern physics attempts, where possible, to replace a many-body problem by a set of one-body problems, for instance by using the method of elementary excitations discussed in Section 1.4. When this can be done as a first approximation, the interaction between these excitations can be treated as a perturbation, with their strength as a natural small parameter.

Another simple example of a many-body system is provided by the non-ideal gas. If the interactions between the atoms or molecules in a gas are small compared to their kinetic energy, the interactions between two, three, four, ... particles can be treated as a perturbation when the partition function is calculated. The result of such a procedure is an equation of state, for a gas at pressure p and temperature T, in the form of the well-known virial series,

$$p/k_B T = a_0 \rho + a_1 \rho^2 + \cdots \tag{5.54}$$

in which $p/k_B T$ is expanded as a power series in the density ρ. The equation of state of an ideal gas will obviously be the zeroth-order approximation. If one restricts one's attention to interactions between pairs of particles only, van der Waals' equation can be obtained as a good approximation.

Density Expansion of Transport Coefficients

The above discussion concerned the equilibrium properties of fluids. We now turn to their nonequilibrium properties.

The oldest and most widely used method of calculating these is based on the Boltzmann transport equation. The advantage of such an approach is that it enables us to express the transport coefficients (viscosity, heat conductivity, diffusion, etc.) which characterize nonequilibrium processes in terms of the parameters that describe the forces of interaction between the particles. Such an approach, based on first principles, is a physicist's dream, but can only be realized in a few cases. Moreover, the usual technique of the Boltzmann equation only takes account of collisions between pairs of particles, so that it only applies to systems that are sufficiently dilute.

Another useful method is to find a relationship between the transport coefficients and the system's macroscopic properties, such as its density, temperature, and concentration, or in other words a virial expansion of the transport coefficients similar to the virial series of Eq. (54) for the equilibrium equation of state. All transport coefficients are proportional to the number density of particles and to their mean free path, while if only pair collisions are taken into account the latter is inversely proportional to the number density. Thus, the usual technique of the Boltzmann equation will lead to transport coefficients that are independent of the system's density. A dependence on this density, and so a virial expansion, will only arise if account is taken of possible collisions between three or more particles. Here, a triple collision means one in which three particles are simultaneously interacting with each other.

One way of taking into account such multiple collisions is to expand the collision operator in the Boltzmann equation as a power series in the density of particles. A subsequent averaging of the Boltzmann equation will then lead to macroscopic equations for the system in which the transport coefficients μ are expressed as a power series in the number density n, or rather in the product of n and s^3, where s is a characteristic molecular size. Thus, we expect that

$$\mu = \mu_0(t) + \mu_1(T)(ns^3) + \mu_2(T)(ns^3)^2 + \cdots, \tag{5.55}$$

where the coefficients $\mu_0, \mu_1, \mu_2, \ldots$, are functions of temperature and are determined by the effects of collision of two, three, four, ... particles, respectively. The similarity of Eq. (55) for the transport coefficients and Eq. (54) for the equilibrium value of $p/k_B T$ is obvious. Moreover, the coefficients a_j in Eq. (54) and $\mu_j(T)$ in Eq. (55) are each determined by the effects of the collisions of $(j + 2)$ particles, and so there must be some connection between them. In order to find such a connection, it is necessary to find expressions for the virial coefficients in terms of the interaction parameters.

However, the problem is not only the unknown connection between the coefficients in Eqs. (54) and (55). It has been discovered in the past ten years that the series expansion of Eq. (55) is wrong, and that the regular power series

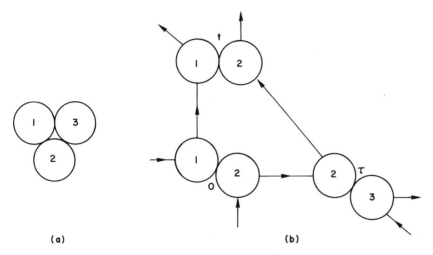

Fig. 1.   Triple collisions: (a) simple; (b) particle 2 collides with particle 1 at time 0 and with particle 3 at time τ; particle 3 deflects it in such a direction that it collides again with particle 1 at time t.

must be replaced by a singular series, thereby breaking the analogy between equilibrium and nonequilibrium properties that seemed so natural. In order to understand the reason for this, let us consider the different forms of multiple collision: we consider first the simplest of these, namely a collision between three particles.

The simplest type of triple collision is that in which three particles are sufficiently close for each one to feel the forces due to the other two. These forces are assumed to be of short range, i.e., of range of order of the molecular size s.† In Fig. 1a, we show a collision of this type for particles with hard-core interactions, i.e., for which the only forces between them occur when they touch each other, and are such as to prevent them from deforming or from penetrating each other. In treatments of the equilibrium properties of a system of particles with such hard-core interactions, this sort of process is usually taken into account by the excluded volume method, which allows for the geometrical restrictions on the arrangement of the particles that arise because they cannot overlap.

Another type of triple collision is depicted in Fig. 1b; here, particles 1 and 2 collide twice, and between these two collisions one of the particles (particle 2 in Fig. 1b) collides with a third particle. Such processes, which are

---

† Such an assumption can be valid, for instance, for systems of neutral atoms or particles interacting via van der Waals forces, but will not be valid in this simple form for charged particles, a point that we return to on p. 216.

called dynamic correlations, are not taken into account at all in the equilibrium theory, where the number of collisions is determined purely geometrically, by combining the volume of the collision cylinders with the number of particles per unit volume. However, these processes are very important for dynamic properties, since they increase the effective time of interaction between the particles that collide twice (particles 1 and 2 in Fig. 1b). Since a quantitative examination of collisions of this type is rather complicated, we will discuss here only some qualitative considerations that explain why such processes lead to the appearance of a singular term in Eq. (55).

The trajectory of particle 3 is a decisive factor in determining whether a second collision will take place between particles 1 and 2. For this to happen, it must collide with particle 2 at some time τ after the first collision, which took place at time 0 say, and send particle 2 in such a direction and at such a speed that it collides again with particle 1, at time t say. In Fig. 2, we show the particles in a coordinate system in which particle 1 is at rest between its two collisions with particle 2. Initially, and so up to the time τ when particle 3 collides with it, particle 2 moves away from particle 1 with velocity v_{21}, and so, as we can see from the figure, particle 3 must deflect it into a solid angle of order $(s/v_{21}\tau)^2$, provided that $v_{21}\tau \gg s$. This process can take place at any time between τ_{coll}, where τ_{coll} is the minimum time required for a collision, and t. Hence, we expect that the volume of phase space available for particle 3 so that it can take part in these dynamic correlations will be proportional to

$$I = \int_{\tau_{coll}}^{t} \left(\frac{s}{v_{21}\tau} \right)^2 d\tau. \qquad (5.56)$$

A rigorous but complicated analysis shows that it is indeed just an integral of this type that appears in the contribution to the coefficient of $\mu_1(t)$ in Eq. (55) from collisions of the type shown in Fig. 1b. In an analogous manner, the coefficient of $\mu_2(T)$, which is the term associated with fourfold collisions, will contain contributions from processes which are the natural generalization of that shown in Fig. 1b. One such process is that in which particle 2 collides with particle 3 and particle 1 with another particle 4 between the times τ_{coll} and t, and another is that in which particle 2 collides with two other particles (particles 3 and 4) in this time interval.

For a three-dimensional gas, the integral I is finite for large values of t. If, however, we consider a hypothetical two-dimensional gas, the solid angle $(s/v_{12}\tau)^2$ in I will be replaced by a planar angle $(s/v_{12}\tau)$, and the corresponding integral will diverge logarithmically for large values of t. For the three-dimensional system, one can show that analogous divergent in-

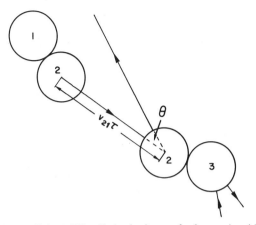

Fig. 2.   The triple collision of Fig. 1b, in the frame of reference in which particle 1 is at rest from time 0 to time t.

tegrals will occur in contributions from dynamic correlations to the coefficients $\mu_2(T)$, etc., of Eq. (55).

In order to complete the calculation, we must eliminate one additional defect of the theory. So far, we have assumed that particles 1 and 2 can move with a velocity of fixed magnitude and direction for an arbitrary long time between their successive collisions. In a real fluid, however, a given particle cannot move with a fixed velocity over more than a few mean free paths, because of random collisions (i.e., not of a type associated with dynamic correlations) with other particles. The natural way to take account of such collisions is to assume that they interrupt the motion of the particles 1 and 2, and so prevent them from undergoing a second collision with each other. It is convenient to define a mean free time τ_f, which is just the mean time taken by a particle to traverse a mean free path, or in other words the mean time between successive collisions of a particle. The effect of the random collisions is to introduce a factor $\exp(-\tau/\tau_f)$ into the integrand of Eq. (56). As a result, the corresponding integral for the two-dimensional system will no longer diverge as $t \to \infty$, but will be proportional to $\ln(\tau_f/\tau_{coll})$. However, in a system of d dimensions, the mean free time, like the mean free path, is proportional to ns^d, where n is the number density of particles. Similar considerations apply to the terms that we found previously to be divergent for a three-dimensional system. Hence, in the three-dimensional case, Eq. (55) will be replaced by one of the form

$$\mu = \mu_0 + \mu_1(T)(ns^3) + \mu_2(T)(ns^3)^2[b_2 \ln(ns^3) + c_2] + \cdots, \quad (5.57)$$

where b_2 and c_2 are numerical constants. For a two-dimensional system, a logarithmic term will appear already in the coefficient of $\mu_1(T)$; for such a system s^3 must be replaced by s^2 throughout the equation, of course. The series (57) is another example of the nonanalytic behavior discussed in Section 4.1, and the physical reason for it is as follows. Strictly speaking, because of the frequent random collisions, we cannot describe nonequilibrium phenomena by a series in which three-particle, four-particle, etc., processes are treated separately and assumed to make progressively smaller contributions as the number of particles involved increases. While the use of a mean free path or time does enable us to use such an expansion, there is no reason to expect this expansion to be an analytic function of the density.

The nonanalyticity appearing in Eq. (57) was first discovered by the numerical analysis of the velocity autocorrelation function for a gas of hard disks. It was found that this function does not decrease exponentially for large values of the time t, as was expected, but instead decreases much less rapidly, namely as $t^{-d/2}$, where d is the dimensionality of the space. This phenomenon is known as the long time tail. As we saw in Section 4.4, integrals of the correlation functions determine the transport coefficients. Hence, the long time tail of the correlation functions is directly related to the nonanalyticity of expansion (57) for the transport coefficients.

Numerical calculations have shown that this long time tail of the correlation functions is analogous to the vortex-type phenomena in hydrodynamics. When a body starts to move through a fluid, a vortex appears round it, because part of the momentum transferred by the body to the fluid elements located in front of it is returned to the body from the elements of fluid located behind it. Such a process leads to an increase in the velocity autocorrelation function, and dynamic correlations of the sort depicted in Fig. 1b are the microscopic basis for this phenomenon.

Low-Density Systems of Charged Particles

In the above problem, our expansion parameter was chosen to be the density, even though it leads to a singular series for the transport coefficients that describe the nonequilibrium properties of the system. Another sort of singularity, namely nonintegral powers, occurs in the treatment of the equilibrium properties of a low-density system of charged particles of one sign moving against a uniform background of charge of opposite sign, when the density is chosen as the expansion parameter, as we now show.

For a system of charged particles, the Coulomb forces between them cannot be treated on the basis of interactions between successively larger number of particles, because of the long-range nature of the electrostatic interactions. Hence, essentially different results are to be expected. If we denote by e_i the charge and by $\mathbf{r}_i^k$ the position of the kth particle of type i, the electrostatic interaction contributes to the energy of the gas a term

$$E_s = \frac{1}{2} \sum_{i,k} \sum_{(j,l) \neq (i,k)} \frac{e_i e_j}{|\mathbf{r}_i^k - \mathbf{r}_j^l|} = \frac{1}{2} \sum_{i,k} e_i \phi(\mathbf{r}_i^k), \qquad (5.58)$$

where $\phi(\mathbf{r}_i^k)$ is the electrostatic potential produced at point $\mathbf{r}_i^k$ by all the charges except the kth one of type i. It is convenient to introduce average variables, and denote by n_{i0} the mean density of particles of type i per unit volume, and by ϕ_i the mean value of $\phi(\mathbf{r}_i^k)$. If we denote the volume of the system by V, we can then write

$$E_s = \frac{1}{2} V \sum_i n_{i0} e_i \phi_i. \qquad (5.59)$$

In order to calculate $\phi(\mathbf{r})$, and hence ϕ_i, we use the same method as that used to derive the Thomas–Fermi equation in Section 2.4. If $n_i(\mathbf{r})$ denotes the density of particles of type i at point $\mathbf{r}$, the charge density at point $\mathbf{r}$ is just $\sum_i n_i(\mathbf{r})e_i$. Hence, one relationship between the potential ϕ and the n_i is provided by Poisson's equation,

$$\nabla^2 \phi = -4\pi \sum_i n_i(\mathbf{r})e_i. \qquad (5.60)$$

A second equation connecting these quantities is provided by the Boltzmann distribution for particles having a given energy, according to which

$$n_i(\mathbf{r}) = n_{i0} \exp(-e_i \phi(\mathbf{r})/k_B T) \approx n_{i0} - n_{i0} e_i \phi(\mathbf{r})/k_B T. \qquad (5.61)$$

In deriving the second equality, we have assumed that $e_i \phi(\mathbf{r})/k_B T$ is a small quantity, i.e., our small parameter is essentially the ratio of the particle's potential energy to its kinetic energy. On combining Eqs. (60) and (61), and making use of the charge neutrality condition $\sum_i n_{i0} e_i = 0$, we find that

$$\nabla^2 \phi = \kappa^2 \phi, \qquad (5.62)$$

where

$$\kappa^2 = 4\pi \sum_i n_{i0} e_i^2 / k_B T. \qquad (5.63)$$

The quantity κ^{-1} is called Debye's screening radius.†

A spherically symmetric solution of Eq. (62), which is expected to be the solution relevant to the potential around a charge at the origin, is

$$\phi(r) = A \exp(-\kappa r)/r. \qquad (5.64)$$

---

† For a degenerate electron gas, the relevant small parameter is the ratio of the potential energy to the Fermi energy E_F, rather than to $k_B T$, and it is found that

$$\kappa^2 = 6\pi \sum_i n_{i0} e_i^2 / E_F. \qquad (5.63a)$$

As $r \to 0$, ϕ must tend to the potential field of an unscreened charge, so that in the neighborhood of a charge of type i we can write

$$\phi(r) = e_i \exp(-\kappa r)/r \approx (e_i/r) - e_i \kappa + \cdots. \tag{5.65}$$

In this expansion, the first term is the potential field of the charge e_i, while the second one is the mean potential due to the other charges, which we denoted by ϕ_i in Eq. (59).† On substituting this value for ϕ_i in Eq. (59), we find that

$$E_s = -\frac{1}{2} V \kappa \sum_i n_{i0} e_i^2 = -V \left(\frac{\pi}{k_B T}\right)^{1/2} \left(\sum_i n_{i0} e_i^2\right)^{3/2}. \tag{5.66}$$

From this equation, a correction to the free energy of the gas, and hence to the equation of state as compared to that of an ideal gas, can be derived by using the well-known thermodynamic relationships. A simple calculation shows that the equation of state becomes

$$\frac{p}{k_B T} = \sum_i n_{i0} - \frac{\sqrt{\pi}}{3} \left(\frac{\sum_i n_{i0} e_i^2}{k_B T}\right)^{3/2}, \tag{5.67}$$

a result known as the Debye–Hückel formula.

We note the following two points in connection with the above analysis. First, the Coulomb potential due to any given charge is screened by the other charges, i.e., the polarization that it produces in the charged gas causes its field to decrease exponentially, with a decay length that is equal to the Debye radius κ^{-1}. Second, while the first term in the virial series (67) for a system of charged particles is just the number density of particles, as for an ideal gas, the second term is proportional to the $\frac{3}{2}$ power of this density, rather than to its square as in the virial series of Eq. (54) for a gas of uncharged particles.

The High-Density Electron Gas

Up to now, we have discussed a low-density system of charged particles, in which the interactions between the particles could be regarded as a small perturbation. Surprisingly enough, it turns out that this interaction can also be treated as a small perturbation for a high-density system of charged particles, such as electrons, interacting according to Coulomb's law. We consider a quantum-mechanical treatment of such a gas of electrons, for

---

† The condition for such an expansion to be valid, $\kappa r \ll 1$, is essentially the same as that for the expansion of Eq. (61) to be valid, i.e., $e_i \phi(r_0)/k_B T \ll 1$ where r_0 is the mean distance between adjacent particles.

which the Hamiltonian has the form

$$H = \frac{-\hbar^2}{2m} \sum_i \nabla_i^2 + \sum_{i \neq j} \frac{e^2}{|\mathbf{r}_i - \mathbf{r}_j|}, \tag{5.68}$$

where $\mathbf{r}_i$ denotes the position coordinate of the ith electron. Since a system of charged particles all of the same kind is unstable, we assume that the electrons move against a background of positive charge spread uniformly throughout the system. A more exact treatment of a system of electrons and ions will be considered in Section 5.3.

In order to obtain a small parameter for our system, we introduce the average distance r_0 between the particles, according to the formula

$$4\pi r_0^3/3 = V/N, \tag{5.69}$$

where V is the volume of the system and N the total number of electrons. On substituting the dimensionless variables

$$\mathbf{s}_i = \mathbf{r}_i/r_0 \tag{5.70}$$

into Eq. (68), we find that

$$H = \frac{\hbar^2}{2mr_0^2}\left[-\sum_i \nabla_{s_i}^2 + 2\left(\frac{e^2 m r_0}{\hbar^2}\right)\sum_{i \neq j} \frac{1}{|\mathbf{s}_i - \mathbf{s}_j|}\right]. \tag{5.71}$$

Since $\hbar^2/me^2 = a_0$, the Bohr radius, which is the natural unit of length for calculations on atoms [cf. Eq. (2.101b), which was in SI units], we can write Eq. (71) in the form

$$\mathsf{H} \equiv \frac{2mr_0^2}{\hbar^2} H = -\sum_i \nabla_{s_i}^2 + 2\lambda \sum_{i \neq j} \frac{1}{|\mathbf{s}_i - \mathbf{s}_j|}, \tag{5.72}$$

where for a high-density system

$$\lambda = r_0/a_0 \ll 1. \tag{5.73}$$

Since r_0 decreases as the density increases, this system becomes more ideal as its density increases, in contrast to a nonideal gas of neutral particles.

The Hamiltonian H of Eq. (72) is of the form $H_0 + \lambda H_1$, and so its eigenvalues and eigenfunctions can in principle be calculated by perturbation theory. However, this calculation is too complicated to be described here. One interesting feature of it is that it proves necessary to take into account the most important terms of all orders of perturbation theory, instead of considering the complete perturbation term up to some definite order. The final result for the mean energy per particle, E/N, is given (in Rydberg units) by

$$E/N = 2.21/\lambda^2 - 0.916/\lambda + 0.062 \ln \lambda - 0.096 + a\lambda + b\lambda \ln \lambda + c\lambda^2 + \cdots. \tag{5.74}$$

The equation of state for the high-density electron gas will involve on its right-hand side terms of the same form as those in Eq. (74), just as for the low density gas E/N and the equation of state [Eqs. (66) and (67)] involved terms of the same form.

We have now considered the electron gas in the limiting cases of both very low ($\lambda \gg 1$) and very high ($\lambda \ll 1$) densities. Unfortunately, the most interesting case for the application of the electron gas model is to real metals, where λ assumes intermediate values, typically $2 \leq \lambda < 5.5$. For such cases, one has to use some sort of interpolation formula. Furthermore, as we will see in Section 5.3, the effect of the background of positive charge can be taken into account more precisely than by the assumption of a uniform background used in the above example. In fact, while for a system containing just one type of particle the only small parameter is the density, for a mixture of two different types of charged particle another small parameter can usually be associated with the ratio of some distinctive property for the two types. For instance, the ratio of the mass of an electron to that of a nucleon, m/M, is approximately $1/1800$, and this ratio can often be used as a small parameter in the treatment of a mixture of electrons and nuclei by perturbation theory. This is the basis of the adiabatic approximation, which is widely used in the theory of atoms, molecules, and solids, and which we consider further in Section 5.3.

Breakdown of Perturbation Theory

In order to apply perturbation theory, we must first make sure that its basis, the assumption that a small perturbation leads to small changes, is not violated. A well-known example where this is not the case is that of superconductivity. For many years, people tried unsuccessfully to use different forms of perturbation theory to explain this phenomenon. The reason that they failed is that an infinitesimally small attractive interaction between Fermi particles, in our example electrons, leads to a basic change in the ground state of the system rather than just to a small shift in its energy. As is now well known, the existence of such an attraction transforms the system from a set of independent, noninteracting particles into a system of electron pairs, the components of which have opposite directions of momentum and of spin, because such a transformation lowers the system's energy. Since this formation of pairs involves a binding energy, an energy gap appears between the system's excited states, in which some pairs are unbound, and its ground state, and this, in turn, explains the phenomenon of superconductivity.

Another case in which a small change produces a large effect is that of a system close to a phase transition. As we have discussed previously, a small change of temperature near the order–disorder transition temperature will have a large effect on the symmetry. Hence, perturbation theory cannot be applied to a discussion of the physical phenomena associated with phase transitions.

Decrease of the Order of a Differential Equation

Another situation in which the term multiplied by a small parameter cannot be treated by simple perturbation theory occurs when this parameter multiplies the highest derivative that appears in a differential equation. A simple example, originally discussed by Prandtl in 1931, is that of a damped harmonic oscillator whose mass m is allowed to tend to zero. The equation of motion of this oscillator,

$$m\ddot{y} + a\dot{y} + Ky = 0, \qquad (5.75)$$

is, of course, another special case of Eq. (1), with $b = 0$ and $g = 0$. We now compare two ways of taking into account the small value of m as $m \to 0$. Either we can just ignore the term $m\ddot{y}$ in the first approximation, or we can solve Eq. (75) exactly and then let $m \to 0$. While the second method is obviously preferable, it is frequently not practicable for more complicated equations, and this is why the first method is very appealing. For instance, in the problem of space-charge-limited current considered in Section 1.5, it is extremely difficult to solve Eq. (1.17) unless the diffusion term, which is the only one involving the second derivative of $n(x)$, is ignored.

If we ignore the term $m\ddot{y}$ in Eq. (75), we decrease the order of the differential equation, and hence the number of boundary conditions that can be imposed on the solution. This is the source of all our subsequent difficulties. The solution of Eq. (75) with $m = 0$ is

$$y_{\text{I}} = y_{\text{I}}(0) \exp(-Kt/a), \qquad (5.76)$$

while the solution of Eq. (75) for small m (i.e., $m \ll a^2/K$) is approximately

$$y_{\text{II}} = A \exp(-Kt/a) + B \exp(-at/m). \qquad (5.77)$$

The second solution y_{II} tends to y_{I} as $m \to 0$, provided that $A = y_{\text{I}}(0)$, only if t does not simultaneously approach zero. However, if we impose on our solution the boundary condition $y(0) = 0$, we see that $y_{\text{I}} = 0$, and bears no relationship to y_{II}, which in this case is given by

$$y_{\text{II}} = A[\exp(-Kt/a) - \exp(-at/m)]. \qquad (5.78)$$

The solution y_{I} which has the same form as y_{II} when $t \to \infty$ is

$$y_{\text{I}} = A \exp(-Kt/a). \qquad (5.79)$$

In Fig. 3 we show y_I/A (the solid line) and y_{II}/A for different small values of m (the broken lines), as given by Eqs. (79) and (78). As $m \to 0$, y_I becomes a good approximation to y_{II} for an increasingly larger range of t, but it is never a good approximation at $t = 0$. Mathematically, the reason for this is that, as $m \to 0$, $m\ddot{y}_{II}(0) \sim -a^2/m$. This is not a small quantity compared to the other terms in Eq. (75), and is in fact of the same order of magnitude as $a\dot{y}(0)$. Thus, the fact that a term is multiplied by a small parameter is not sufficient on its own to ensure that the term is small, and so can be treated by perturbation theory. However, as we can see from the figure, if we are satisfied with a solution that is a good approximation for a large range of values of t, i.e., those for which $m\ddot{y}_{II}$ is indeed small, the first method is adequate.

A similar problem arises in hydrodynamics, where the equations of motion for an ideal fluid and a viscous one are of different orders. As a result, even a fluid of low viscosity cannot be approximated by an ideal fluid in the region close to the boundary of the solid, because an ideal fluid cannot satisfy the boundary conditions required for one whose viscosity is nonzero.

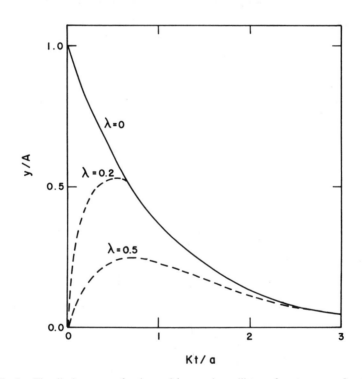

Fig. 3. The displacement of a damped harmonic oscillator of mass m, as a function of time, for three different values of the small parameter $\lambda = mK/a^2$.

As we saw in Section 1.6, this sort of problem must be treated by the method of boundary layers. In the first part of Section 5.4, we compare various ways of treating equations in which a small parameter multiplies the highest derivative.

5.2   Integral Equation Formulations of Perturbation Theory

The use of infinite series to solve problems that cannot readily be solved by other methods is a very widespread and valuable technique. The approach to perturbation theory that we described in Section 5.1 was based on the simplest possible version of this technique, according to which the quantities of interest are expressed as power series in λ, which are assumed to converge, and the coefficients of corresponding powers of λ on each side of the equation are then equated. However, numerous more sophisticated versions of perturbation theory exist. In this section, we will not try to describe and discuss all of these, most of which are treated at length in numerous specialist texts as well as in textbooks on quantum mechanics and statistical mechanics. Rather, in accordance with the aims and scope of this book, we will consider the relationship between some of the main methods that are in use, restricting our attention in each case to the simplest examples.

Integral Equations

A useful alternative to the straightforward substitution technique discussed in Section 5.1 is to formulate the problem as an integral equation, rather than as a differential equation. This is a very powerful standard technique in the study of approximate solutions because if two functions are approximately equal, integrals of them over a given interval will always be close to each other, but their differentials need not be. For instance, $f_0(x) = 1$ is a moderate approximation to $f(x) = 1 + 0.1 \sin x$, and $\int_0^\pi f_0(x)\, dx$ is quite close to $\int_0^\pi f(x)\, dx$, but $f_0'(x) = 0$ is not a good approximation to $f'(x) = 0.1 \cos x$.

We apply this technique first to the problem

$$m\, d^2y/dt^2 + Ky = \phi(t) - \lambda(a\, dy/dt + bf(t)y), \qquad y(0) = p, \qquad \dot{y}(0) = q. \tag{5.80}$$

This is just the problem of Eqs. (1) and (2), with a parameter λ added as in Eq. (5). The right-hand side of Eq. (80) can be treated as an inhomogeneous term, similar to $\phi_k(t)$ in Eq. (14). Thus, in view of Eq. (15),

$$y(t) = p \cos \omega_0 t + \frac{q}{\omega_0} \sin \omega_0 t$$

$$+ \frac{1}{m\omega_0} \int_0^t \left[\phi(s) - \lambda a \frac{dy}{ds} - \lambda b f(s) y(s) \right] \sin[\omega_0(t - s)] \, ds. \quad (5.81)$$

On performing a partial integration for the term involving dy/ds, we find that Eq. (81) can be written in the form

$$y(t) = h(t) + \lambda \int_0^t K(t, s) y(s) \, ds, \quad (5.82)$$

where

$$h(t) = p \cos \omega_0 t + \frac{q}{\omega_0} \sin \omega_0 t + \frac{1}{m\omega_0} \int_0^t \phi(s) \sin[\omega_0(t - s)] \, ds$$

$$+ \frac{\lambda a}{m\omega_0} p \sin \omega_0 t \quad (5.83)$$

and

$$K(t, s) = -(a/m) \cos[\omega_0(t - s)] - b f(s) \sin[\omega_0(t - s)]. \quad (5.84)$$

Equation (82) is just a Volterra integral equation of the second kind. To find $y(t)$, we use the technique of repeated substitution or iteration, denoting by $y^{(n)}$ the nth approximation to y, and writing

$$y^{(n)}(t) = h(t) + \lambda \int_0^t K(t, s) y^{(n-1)}(s) \, ds. \quad (5.85)$$

It is convenient to define functions $v_n(t)$ by means of

$$v_0(t) = y^{(0)} = h(t), \qquad v_n(t) = y^{(n)}(t) - y^{(n-1)}(t), \quad (5.86)$$

so that

$$v_n(t) = \lambda \int_0^t K(t, s) v_{n-1}(s) \, ds. \quad (5.87)$$

The above result is exactly the same as that obtained for this problem in Section 5.1, with $v_n = \lambda^n y_n$. However, the formulation of the result in terms of a set of integral equations enables us to derive valuable information about the convergence of the series $y = \sum_0^\infty \lambda^n y_n = \sum_0^\infty v_n$. If $f(t)$ and $\int_0^t |\phi(s)| \, ds$ are bounded for all t, so are $h(t)$ and $K(t, s)$ for all t and s, i.e., we can define C and M such that

$$|h(t)| \leq C, \qquad |K(t, s)| \leq M. \tag{5.88}$$

Then, from Eqs. (86) and (87),

$$|v_1(t)| \leq |\lambda| CMt, \tag{5.89a}$$

$$|v_2(t)| \leq |\lambda| \int_0^t M|\lambda| CMt \, dt = \frac{C(M|\lambda|)^2 t^2}{2!} \tag{5.89b}$$

and, by induction,

$$|v_n(t)| \leq C(M|\lambda|)^n t^n/n!. \tag{5.89c}$$

Thus,

$$\sum_{n=0}^{\infty} |v_n(t)| \leq C \exp(M|\lambda|t), \tag{5.90}$$

and so the perturbation series converges for all values of λ and t. Incidentally, Eqs. (89) and (90) enable us to derive an upper bound to the error involved in terminating the perturbation series after n terms.

For the anharmonic oscillator described by Eq. (21), we can still convert the problem into an integral equation, which in this case will be

$$y(t) = p \cos \omega_0 t - \frac{c}{m\omega_0} \int_0^t [y(s)]^3 \sin[\omega_0(t - s)] \, ds. \tag{5.91}$$

However, this is a nonlinear equation, so that if we define an iterative solution by Eq. (85), the relationship between successive approximations is no longer given by Eqs. (86) and (87). Thus, our conclusions about convergence of the perturbation series cannot readily be applied to this problem.

Green's Functions

In order to convert into an integral equation a problem such as that of Eqs. (1) and (3), with homogeneous boundary conditions at two points, it is convenient to introduce the concept of Green's functions. The Green's function $G(\mathbf{r}, \mathbf{r}', E)$ of a linear operator $H(r)$ is defined to be such that the solution of

$$(H(\mathbf{r}) - E)y(\mathbf{r}) + f(\mathbf{r}) = 0, \tag{5.92}$$

satisfying the relevant boundary conditions, is

$$y(\mathbf{r}) = \int_\Omega G(\mathbf{r}, \mathbf{r}', E) f(\mathbf{r}') \, dv', \tag{5.93}$$

where Ω is the distance between the boundaries for a one-dimensional problem, or the volume for a three-dimensional one. In particular, if we choose $f(\mathbf{r}') = \delta(\mathbf{r}' - \mathbf{r}_0)$, we see that $G(\mathbf{r}, \mathbf{r}_0, E)$ satisfies the equation

$$(H(\mathbf{r}) - E)G(\mathbf{r}, \mathbf{r}_0, E) = -\delta(\mathbf{r} - \mathbf{r}_0). \qquad (5.94)$$

If we denote by G_0 the Green's function of the operator H_0, we find that the differential equation of Brillouin–Wigner perturbation theory

$$(H_0 - E)y = -\lambda H_1 y \qquad (5.95)$$

can be replaced by the integral equation

$$y(\mathbf{r}) = \lambda \int_\Omega G_0(\mathbf{r}, \mathbf{r}', E)H_1(\mathbf{r}')y(\mathbf{r}') \, dv'. \qquad (5.96)$$

Similarly, Eq. (47) is replaced by the integral equation for $y - y_0$,

$$y(\mathbf{r}) - y_0(\mathbf{r}) = \int_\Omega G_0(\mathbf{r}, \mathbf{r}', E)[E_0 + \lambda H_1(\mathbf{r}') - E]y_0(\mathbf{r}') \, dv'$$

$$+ \lambda \int_\Omega G_0(\mathbf{r}, \mathbf{r}', E)H_1(\mathbf{r}')[y(\mathbf{r}') - y_0(\mathbf{r}')] \, dv'. \qquad (5.97)$$

Equations (96) and (97) involve definite integrals, in contrast to Eq. (82). Such equations are known as Fredholm integral equations, of the first and second kind, respectively.

Brillouin–Wigner and Rayleigh–Schrödinger Perturbation Theory

In order to make contact with the perturbation theory described in Section 5.1, we now express the Green's function $G_0(\mathbf{r}, \mathbf{r}', E)$ in terms of the eigenfunctions u_k of H_0; we assume that H_0 is a Hermitian operator, so that the functions u_k can be chosen to be orthonormal, and that these functions form a complete, denumerable set. On using Eq. (92) and definition (93) of the Green's function, and expressing y and f as linear combinations of the functions u_k, we readily find that

$$G_0(\mathbf{r}, \mathbf{r}'; E) = \sum_{k=0}^{\infty} \frac{u_k^*(\mathbf{r}')u_k(\mathbf{r})}{(E - E_k)}. \qquad (5.98)$$

We now substitute this expression, together with

$$y = \sum_n c_n u_n \qquad (5.99)$$

into Eq. (96), which is the basis of Brillouin–Wigner perturbation theory. The resulting equation is then multiplied by $u_m^*(\mathbf{r})$ and integrated over Ω, in order to obtain the equation

$$c_m = \lambda \sum_{n=0}^{\infty} \frac{c_n H_{mn}^1}{(E - E_m)},$$ (5.100)

Since, from Eqs. (7), (37), and (99),

$$c_m = \sum_{j=0}^{\infty} \lambda^j c_m^j,$$ (5.101)

on equating corresponding powers of λ in Eq. (100) for $m \neq 0$ we once again obtain Eq. (51). In order to derive Eq. (50), we multiply Eq. (100) for $m = 0$ by $(E - E_0)$, set $c_0 = 1$, and substitute $E = \Sigma_n \lambda^n \varepsilon_n$. Thus, the eigenvalue expansion of G_0, Eq. (98), leads to exactly the same results as previously. However, one advantage of the Green's function method is that different expressions for G_0 can often conveniently be used, as in the example from scattering theory that we consider below.

If we attempt to use a Green's function G_0 to convert the basic equation of Rayleigh–Schrödinger perturbation theory, Eq. (31), into an integral equation, we are immediately confronted with a problem. The Green's function that we require, $G_0(\mathbf{r}, \mathbf{r}', E_0)$, is infinite, because of the term with $k = 0$ in the summation in Eq. (98). This singularity of the Green's function at an eigenvalue of the operator follows immediately from the basic definition of the Green's function, Eq. (93), if H is a Hermitian operator. For, in that case, on multiplying equation (92) by an arbitrary function $u^*(\mathbf{r})$, integrating over Ω, and using the fact that H is Hermitian, we find that

$$\int_{\Omega} y^*(\mathbf{r})(H - E)u(\mathbf{r}) \, dv = - \int u^*(\mathbf{r}) f(\mathbf{r}) \, dv.$$ (5.102)

If E is an eigenvalue of H, and we choose $u(\mathbf{r})$ to be the corresponding eigenfunction, the left-hand side of this equation vanishes, while the right-hand side does not do so if we choose, for instance, $f(\mathbf{r}) = u(\mathbf{r})$. Since a solution of Eq. (92) always exists if the Green's function does, we conclude that $G(\mathbf{r}, \mathbf{r}', E)$ must have a singularity at each eigenvalue of H. We see from Eq. (98) that this singularity is in fact a pole, whose order equals the degeneracy of the eigenvalue.

It is often very inconvenient to use functions with singularities for real values of E. One way to overcome this problem is to consider $G(\mathbf{r}, \mathbf{r}', E \pm i\delta)$, where δ is a small quantity that is allowed to tend to zero at the end of the calculation. If we use this method and consider Eq. (31) with E_0 replaced by $E_0 + i\delta$, we readily obtain the results of Rayleigh–Schrödinger perturbation theory. In particular, Eq. (40) for ε_j follows from the requirement that the right-hand side of Eq. (31), which is analogous to $f(\mathbf{r})$ in Eq. (102), be orthogonal to $y_0(\mathbf{r})$, so that the equations are satisfied when the limit $\delta \to 0$ is taken in $G_0(\mathbf{r}, \mathbf{r}', E_0 + i\delta)$.

*Convergence of the Perturbation Series

If we set

$$w(\mathbf{r}) = y(\mathbf{r}) - y_0(\mathbf{r}), \qquad h(\mathbf{r}) = \int_\Omega G_0(\mathbf{r}, \mathbf{r}', E)(E_0 + \lambda H_1(\mathbf{r}') - E)y_0(\mathbf{r}') \, dv',$$

$$(5.103)$$

Eq. (97) can be expressed in the form

$$w(\mathbf{r}) = h(\mathbf{r}) + \lambda \int_\Omega G_0(\mathbf{r}, \mathbf{r}', E)H_1(\mathbf{r}')w(\mathbf{r}') \, dv'. \qquad (5.104)$$

The simplest method of solving such an integral equation is by iteration, just as in our previous example. The zeroth-order approximation is $w(\mathbf{r}) = h(\mathbf{r})$, and at each subsequent step the form of $w(\mathbf{r})$ found at the previous step is substituted on the right-hand side of Eq. (104). This process is completely equivalent to expanding y as a power series in λ, i.e., to the perturbation series derived previously. Hence, the conditions for the convergence of the Brillouin–Wigner perturbation series are the same as those for the convergence of the iterative solution of Eq. (104). The resulting series is known in the theory of integral equations as the Neumann series.

However, the above method contains one major pitfall. Both $h(\mathbf{r})$ and $G_0(\mathbf{r}, \mathbf{r}', E)$ are functions of E, while the convergence of the iterative method can be discussed readily only if they are fixed. Hence, the value of E, which is one of the quantities that we aim to calculate, must be fixed before we start to construct the Neumann series. While this difficulty can be overcome, for instance, by establishing conditions for the series convergence at all values of E in the range of interest, it is too complicated a problem to be discussed here. A similar situation arises with Rayleigh–Schrödinger perturbation theory. The only simple result that we can derive is as follows. If, for a given λ, $H_0 + \lambda H_1$ has an eigenvalue equal to one of those of H_0, say E_1, the iteration method will not converge, since $G_0(\mathbf{r}, \mathbf{r}', E_1)$ is infinite. Hence, the radius of convergence of the perturbation series, regarded as a power series in λ, cannot exceed the smallest value of $|\lambda|$ for which $H_0 + \lambda H_1$ has such an eigenvalue. However, the series may well not converge for smaller values of λ than this.

Scattering Theory—The First Born Approximation

As we mentioned above, it is not essential to express the Green's function G_0 in terms of the eigenfunctions of H_0, and such a representation is not necessarily the most convenient one. For instance, an explicit form of the Green's function can readily be found for the Sturm–Liouville operator, which is of the form $H_0 = (d/dx)[p(x)\,d/dx] + q(x)$. The method of doing this and the application of the resulting G_0 to a simple problem are discussed in Problems 5.5 and 5.6. Another operator for which the Green's function is known explicitly, at least for certain types of boundary conditions, is the Laplacian ∇^2. For instance, the solution of the three-dimensional Helmholtz equation

$$(\nabla^2 - E)G_0(\mathbf{r}, \mathbf{r}', E) = -\delta(\mathbf{r} - \mathbf{r}') \tag{5.105}$$

that corresponds to an outgoing wave as $|\mathbf{r}| \to \infty$ is

$$G_0(\mathbf{r}, \mathbf{r}', E) = \exp(ik|\mathbf{r} - \mathbf{r}'|)/(4\pi|\mathbf{r} - \mathbf{r}'|), \quad \text{where} \quad k^2 = -E. \tag{5.106}$$

An example of the application of the above Green's function is provided by the quantum-mechanical treatment of the scattering of a particle incident with wave function $\psi_0 = \exp(i\mathbf{k}_0 \cdot \mathbf{r})$ by a body that produces a potential $V(\mathbf{r})$, a problem that we considered in Sections 4.1 and 4.2. Schrödinger's equation can be written in the form

$$(\nabla^2 + k^2)\psi(\mathbf{r}) = (2m/\hbar^2)V(\mathbf{r})\psi(\mathbf{r}). \tag{5.107}$$

On using the Green's function G_0, Eq. (106), we see that the solution corresponding to the incoming wave plus a scattered wave is just

$$\psi(\mathbf{r}) = \exp(i\mathbf{k}_0 \cdot \mathbf{r}) - \frac{m}{2\pi\hbar^2} \int \frac{V(\mathbf{r}')\psi(\mathbf{r}')}{|\mathbf{r} - \mathbf{r}'|} \exp(ik|\mathbf{r} - \mathbf{r}'|)\,dv'. \tag{5.108}$$

The asymptotic form of this for large $|\mathbf{r} - \mathbf{r}'|$ is as in Eq. (4.48). For an iterative solution, we use $\psi = \psi_0$ as our zeroth-order approximation, and substitute this in the integral on the right-hand side of Eq. (108). The resultant expression for ψ is known as the first Born approximation.

Dyson's Equation

So far, we have only used the Green's function G_0 of H_0 to find solutions y of the homogeneous equation $(H - E)y = 0$. However, there is no reason in principle why we cannot also use it to calculate the Green's function $G(\mathbf{r}, \mathbf{r}', E)$ of the operator H, as defined by Eq. (94). One very important property of G is that its poles at the eigenvalues of H, must as those of

G_0 did at the eigenvalues of H_0. Thus, by locating the poles of G, we can find the eigenvalues of H; an extension of this, considered in Problem 5.8, relates the density of states for a Hamiltonian with a continuous spectrum of eigenvalues to the Green's function G. Another important application of G is in calculating the response of a system having Hamiltonian H to an extra infinitesimal term in the Hamiltonian; this response determines, of course, the system's generalized susceptibility, a property that was discussed at length in Chapter 4.

Since the formal definition of G by Eq. (94) is so similar to that of y by Eq. (92), it is natural to attempt to solve it by the same techniques. Thus we write

$$(H_0(\mathbf{r}) - E)G(\mathbf{r}, \mathbf{r}', E) = -\delta(\mathbf{r} - \mathbf{r}') - \lambda H_1(\mathbf{r})G(\mathbf{r}, \mathbf{r}', E) \qquad (5.109)$$

and solve this equation by means of the Green's function G_0 to obtain

$$G(\mathbf{r}, \mathbf{r}', E) = G_0(\mathbf{r}, \mathbf{r}', E) + \lambda \int G_0(\mathbf{r}, \mathbf{r}'', E)H_1(\mathbf{r}'')G(\mathbf{r}'', \mathbf{r}', E)\, dv''. \qquad (5.110)$$

This equation can be written formally as

$$G = G_0 + \lambda G_0 H_1 G, \qquad (5.111)$$

where the operator H_1 involves integration over $\mathbf{r}''$. Such a form, which is known as Dyson's equation, is obviously well suited to solution by repeated substitution. The results of such an iteration procedure is an infinite series for G,

$$G = G_0 + \lambda G_0 H_1 G_0 + \lambda^2 G_0 H_1 G_0 H_1 G_0 + \cdots \qquad (5.112)$$

with the same convergence problems as for the Brillouin–Wigner perturbation series.

While Dyson's equation is formally very neat, this fact does not necessarily make it any easier to solve. If H_1 were a number, rather than an operator, the series of Eq. (112) could be summed (when convergent) to give $G = G_0/(1 - \lambda H_1 G_0)$. This result also follows from Eq. (111) even if the series does not converge. Unfortunately, however, H_1 is generally an operator, whose expectation value depends on the state under consideration. In order to obtain an algebraic equation, we can express G in terms of the eigenfunctions of H_0, using the expansion of equation (98) for G_0. We then find, on substituting this in equation (112), that

$$G(\mathbf{r}, \mathbf{r}', E) = \sum_n \frac{u_n(\mathbf{r})u_n^*(\mathbf{r}')}{E - E_n} + \lambda \sum_n \sum_m H_{nm}^1 \frac{u_n(\mathbf{r})u_m^*(\mathbf{r}')}{(E - E_n)(E - E_m)} + \cdots. \qquad (5.113)$$

It is only practicable to extend this series to terms of higher order in λ if the vast majority of the matrix elements H_{nm}^1 are zero. In that case, a number of techniques have been developed for summing the series to higher order, and some terms of it to infinite order, with the diagrams introduced by Feynmann being used to identify which terms are included in any partial sum. These techniques are discussed fully in numerous books and articles, and we will not consider them here. Instead, we will just consider two problems that can be solved simply by Dyson's equation.

A trivial, but instructive, example of the use of Dyson's equation is provided by a system in which $H_{mn}^1 = a_n \delta_{m,n}$. In that case, we can sum the perturbation series (112) to infinite order, and find that

$$G(\mathbf{r}, \mathbf{r}', E) = \sum_n \frac{u_n(\mathbf{r})u_n^*(\mathbf{r}')}{E - E_n} \left(1 + \frac{\lambda a_n}{E - E_n} + \frac{\lambda^2 a_n^2}{(E - E_n)^2} \cdots \right)$$

$$= \sum_n \frac{u_n(\mathbf{r})u_n^*(\mathbf{r}')}{E - E_n - \lambda a_n}. \tag{5.114}$$

This result is exact, of course, but is also trivial, since with this form of H_{mn}^1 each eigenfunction u_n of H_0 is also an eigenfunction of H, but with eigenvalue $E_n + \lambda a_n$.

A more interesting example of the application of Dyson's equation is provided by a system containing a single, strongly localized impurity at $\mathbf{r}_0$, so that we can write

$$H_1(\mathbf{r}) = v\delta(\mathbf{r} - \mathbf{r}_0). \tag{5.115}$$

Since, with this potential, $H_{nm}^1 \neq 0$ in general, it is difficult to calculate the wave functions and energies of the perturbed system to more than second order in v by means of Rayleigh–Schrödinger or Brillouin–Wigner perturbation theory. However, if we substitute this perturbation in Dyson's equation, Eq. (110), we find that

$$G(\mathbf{r}, \mathbf{r}', E) = G_0(\mathbf{r}, \mathbf{r}', E) + G_0(\mathbf{r}, \mathbf{r}_0, E)vG(\mathbf{r}_0, \mathbf{r}', E). \tag{5.116}$$

While such an equation can be solved by iteration, it can also be solved explicitly, a process that is equivalent to summing the perturbation series to infinite order. To find the explicit solution, we first of all set $\mathbf{r} = \mathbf{r}_0$ in Eq. (116), and obtain

$$G(\mathbf{r}_0, \mathbf{r}', E) = G_0(\mathbf{r}_0, \mathbf{r}', E) + G_0(\mathbf{r}_0, \mathbf{r}_0, E)vG(\mathbf{r}_0, \mathbf{r}', E). \tag{5.117}$$

Hence,

$$G(\mathbf{r}_0, \mathbf{r}', E) = G_0(\mathbf{r}_0, \mathbf{r}', E)/[1 - vG_0(\mathbf{r}_0, \mathbf{r}_0, E)], \tag{5.118}$$

and on substituting this expression in Eq. (116) we obtain

$$G(\mathbf{r}, \mathbf{r}', E) = G_0(\mathbf{r}, \mathbf{r}', E) + G_0(\mathbf{r}, \mathbf{r}_0, E)vG_0(\mathbf{r}_0, \mathbf{r}', E)/[1 - vG_0(\mathbf{r}_0, \mathbf{r}_0, E)]. \tag{5.119}$$

Incidentally, this result is correct even if $vG_0(\mathbf{r}_0, \mathbf{r}_0, E) > 1$, in which case the perturbation series does not converge.

5.3   Choice of the Small Parameter

In most practical applications of perturbation theory, we want the first two or three terms of the perturbation series to provide a good approximation to the exact solution. A necessary condition for this to happen, as we saw in Section 5.1, is that this series should be an expansion of the solution in terms of increasing powers (not necessarily integral—cf. Eq. (66), for instance) of some small parameter. The choice of this parameter is not always a trivial problem. Even when it is obvious to what it must be proportional, the exact form of the small parameter is not always immediately apparent. For instance, for the anharmonic oscillator of Eq. (21), the expansion parameter is the coefficient of the anharmonic term per unit mass, multiplied by the square of the initial displacement and divided by the square of the natural frequency, i.e., $cp^2/m\omega_0^2$. We note that this parameter is just a dimensionless constant involving the small coefficient and the natural variables of the problem.

Another case, to which we devote most of this section, is that of a system containing heavy and light electrical charges. While the ratio of their masses is a natural dimensionless parameter, it is not trivial to calculate what powers of it enter the perturbation expansion. Moreover, the application of perturbation theory in this case is not as straightforward as in the examples considered so far in this chapter. Another type of problem in which difficulties can arise is one that involves more than one small parameter, a point that we discuss in the last part of this section.

*Quantum-Mechanical Description of a System of Nuclei and Electrons

One system for which the choice of the small parameter and subsequent application of perturbation theory is not immediately obvious is that of a set of positively and negatively charged particles interacting through electrostatic forces. This is one of the most important many-body problems, because it is just these forces that are responsible for the existence of atoms, molecules, and solids, while the simple form of these forces makes it an attractive problem to consider. We restrict our attention to a system of atomic nuclei and electrons, and neglect the effects of spin, magnetic forces, etc. Thus, we write Schrödinger's equation in the form

$$[H_e + H_N + V(\mathbf{r}, \mathbf{R})]\Psi(\mathbf{r}, \mathbf{R}) = E\Psi(\mathbf{r}, \mathbf{R}), \tag{5.120}$$

where $\mathbf{r} = \{\mathbf{r}_i\}$ and $\mathbf{R} = \{\mathbf{R}_a\}$ are the sets of electronic and nuclear coordinates, respectively. Here, H_e and H_N are the kinetic energy operators for electrons (all of mass m) and nuclei (with the nucleus at $\mathbf{R}_a$ having mass M_a), so that

$$H_e = -\frac{\hbar^2}{2m}\sum_i \nabla_i^2, \qquad H_N = -\frac{\hbar^2}{2}\sum_a \frac{\nabla_a^2}{M_a}. \tag{5.121}$$

The potential energy $V(\mathbf{r}, \mathbf{R})$ is just that due to electrostatic forces, and so if the nucleus at $\mathbf{R}_a$ has charge $Z_a e$

$$V(\mathbf{r}, \mathbf{R}) = \sum_{a \neq b} \frac{Z_a Z_b e^2}{|\mathbf{R}_a - \mathbf{R}_b|} + \sum_{i \neq j} \frac{e^2}{|\mathbf{r}_i - \mathbf{r}_j|} - \sum_{i,a} \frac{Z_a e^2}{|\mathbf{r}_i - \mathbf{R}_a|}. \tag{5.122}$$

The wave function $\Psi(\mathbf{r}, \mathbf{R})$, defined by Eq. (120) with the appropriate boundary conditions, contains all the information, for instance, about the properties of solids at the absolute zero of temperature. Unfortunately, however, a macroscopic piece of material contains some 10^{23} particles, and the solution of an equation containing such a large number of variables is beyond the capability of any computer. Moreover, difficulties of principle exist with regard to the calculation of observables from such a cumbersome wave function, the definition of initial condition for nonstationary problems in which $E\Psi$ on the right-hand side of Eq. (120) is replaced by $i\hbar\,\partial\Psi/\partial t$, and so on. This is why approximate methods of solution, such as the method of elementary excitations considered in Section 1.4, are of such great importance. We will consider here the Born–Oppenheimer or adiabatic approximation, originally proposed in 1927, which can in principle be applied to atoms, molecules, and solids. Although more sophisticated methods have been developed in the 50 years that have elapsed since it was proposed, the adiabatic approximation has retained its major importance as a simple and useful method.

The basis for the simplification of the Hamiltonian of Eq. (120) is the great difference between the masses of electrons and nuclei. In the zeroth-order approximation, we can regard the nuclei as having infinite masses ($M_a = \infty$) and so being immovable. Schrödinger's equation then takes the simpler form

$$H_0 \psi(\mathbf{r}; \mathbf{R}) = W(\mathbf{R})\psi(\mathbf{r}; \mathbf{R}), \qquad H_0 = -\frac{\hbar^2}{2m}\sum_i \nabla_i^2 + V(\mathbf{r}, \mathbf{R}), \tag{5.123}$$

where the notation $\psi(\mathbf{r}; \mathbf{R})$ and $W(\mathbf{R})$ means that the electronic wave functions ψ and their energies W depend on the fixed nuclear coordinates $\mathbf{R}$ as parameters. We note that, with this definition, W includes the mutual potential energy of the nuclei. Let us now take into account the fact that the nuclei are not infinitely massive and so can move, but because of their large masses they will experience only small displacements about their equilibrium positions $\mathbf{R}_0$. Thus, we write

$$\mathbf{R} = \mathbf{R}_0 + \alpha\mathbf{u}, \tag{5.124}$$

where the small parameter α equals some power of the ratio of the electron's mass to a typical nuclear mass M_0, i.e.,

$$\alpha = (m/M_0)^p, \tag{5.125}$$

where p is a parameter that depends on the physical approximation used.

The motion of the nuclei leads to the appearance of two kinds of perturbation to the Hamiltonian H_0 of Eq. (123). One of these is associated with the expansion of $V(\mathbf{r}, \mathbf{R})$ and $\psi(\mathbf{r}; \mathbf{R})$ in terms of the small parameter α according to Eqs. (122)–(124), while the other is related to the kinetic energy of the nuclei. The expansion of H_0 in powers of α has the form

$$H_0 = H_0^{(0)} + \alpha H_0^{(1)} + \alpha^2 H_0^{(2)} + \cdots, \tag{5.126}$$

where $H_0^{(k)}$ is an operator acting on the electron coordinates $\mathbf{r}$ and proportional to u^k. The operator H_N that represents the kinetic energy of the nuclei can be rewritten as

$$H_N = \alpha^{-2+1/p} \frac{-\hbar^2}{2m} \sum_a \frac{M_0}{M_a} \frac{\partial^2}{\partial \mathbf{u}_a^2}, \tag{5.127}$$

since $\nabla_a = \partial/\partial\mathbf{R}_a = (1/\alpha)\,\partial/\partial\mathbf{u}_a$. As $(M_0/M_a) \sim 1$, the Hamiltonian H_N is smaller than H_0 by the factor $\alpha^{-2+1/p}$, and not by the factor (m/M_0) as one might have expected. If we wish to restrict our attention to the harmonic approximation in the motion of the nuclei, i.e., the term $\alpha^2 H_0^{(2)}$ in Eq. (126), it is natural for H_N to be of the same order of magnitude as this term. Thus, we must put $-2 + 1/p = 2$, or $p = \frac{1}{4}$, so that the small parameter of this theory is

$$\alpha = (m/M_0)^{1/4}. \tag{5.128}$$

If we now regard $\alpha H_0^{(1)}$, $\alpha^2 H_0^{(2)}$, and $H_N \equiv \alpha^2 H_N^{(2)}$ as perturbations to the zeroth-order Hamiltonian $H_0^{(0)}$, we obtain for the nth energy level the following equations in the zeroth, first, and second orders of perturbation theory [cf. Eq. (10)]:

$$(H_0^{(0)} - E_n^{(0)})\Psi_n^{(0)} = 0 \tag{5.129a}$$

$$(H_0^{(0)} - E_n^{(0)})\Psi_n^{(1)} = -H_0^{(1)}\Psi_n^{(0)} \tag{5.129b}$$

$$(H_0^{(0)} - E_n^{(0)})\Psi_n^{(2)} = -H_0^{(1)}\Psi_n^{(1)} - (H_0^{(2)} + H_N^{(2)} - E_n^{(2)})\Psi_n^{(0)}. \tag{5.129c}$$

Here, we have written

$$E_n = E_n^{(0)} + \alpha E_n^{(1)} + \alpha^2 E_n^{(2)} + \cdots, \tag{5.130}$$

and we have made use of the fact that the first-order correction to the energy $E_n^{(1)}$ is equal to zero. This is the case because we choose $\mathbf{R}_0$, for each state n, to be the equilibrium configuration of nuclei for that state so that $E_n^{(1)} = \Sigma_a \mathbf{u}_a \cdot (\partial E / \partial \mathbf{R}_a) = 0$.

We now proceed to solve the set of Eqs. (129). Let the function $\psi_n^{(0)}(\mathbf{r}; \mathbf{R}_0)$ be the solution of Eq. (129a). Since this is a homogeneous equation, and the operators $H_0^{(k)}$ act only on the electron coordinates,† we can multiply this solution by an arbitrary function $\chi_n^{(0)}(\mathbf{R})$ of the nuclear coordinates. Thus, we write

$$\Psi_n^{(0)}(\mathbf{r}, \mathbf{R}) = \chi_n^0(\mathbf{R})\psi_n^{(0)}(\mathbf{r}; \mathbf{R}_0). \tag{5.131}$$

As we shall see later, the function $\chi_n^{(0)}(\mathbf{R})$ is determined by the higher-order perturbation equations. The equation from first-order perturbation theory, (129b), is an inhomogeneous linear equation, whose solution is a sum of the solution of the homogeneous equation (129a) [multiplied in general by a different function $\chi_n^{(1)}(\mathbf{R})$ of the nuclear coordinates] and a particular solution of the inhomogeneous equation, i.e.,

$$\Psi_n^{(1)}(\mathbf{r}, \mathbf{R}) = \chi_n^{(0)}(\mathbf{R})\psi_n^{(1)}(\mathbf{r}; \mathbf{R}) + \chi_n^{(1)}(\mathbf{R})\psi_n^{(0)}(\mathbf{r}; \mathbf{R}_0). \tag{5.132}$$

On substituting from Eqs. (131) and (132) in the equation for the second-order perturbation, (129c), we find that

$$(H_0^{(0)} - E_n^{(0)})\Psi_n^{(2)} = -H_0^{(1)}[\chi_n^{(0)}(\mathbf{R})\psi_n^{(1)}(\mathbf{r}; \mathbf{R}) + \chi_n^{(1)}(\mathbf{R})\psi_n^{(0)}(\mathbf{r}; \mathbf{R}_0)]$$
$$-[H_0^{(2)} + H_N^{(2)} - E_n^{(2)}]\chi_n^{(0)}(\mathbf{R})\psi_n^{(0)}(\mathbf{r}; \mathbf{R}_0). \tag{5.133}$$

We now compare these equations with those appropriate to perturbation theory for the system in the absence of the term H_N, i.e., where the kinetic energy of the nuclei is neglected. In that case, the nuclear coordinates $\mathbf{R}$ enter as a parameter in the solution at all orders, and we can write

$$(H_0^{(0)} - E_n^{(0)})\psi_n^{(0)}(\mathbf{r}; \mathbf{R}_0) = 0, \tag{5.134a}$$

$$(H_0^{(0)} - E_n^{(0)})\psi_n^{(1)}(\mathbf{r}; \mathbf{R}) = -H_0^{(1)}\psi_n^{(0)}(\mathbf{r}; \mathbf{R}_0), \tag{5.134b}$$

$$(H_0^{(0)} - E_n^{(0)})\psi_n^{(2)}(\mathbf{r}; \mathbf{R}) = -H_0^{(1)}\psi_n^{(1)}(\mathbf{r}; \mathbf{R}) - (H_0^{(2)} - W_n^{(2)})\psi_n^{(0)}(\mathbf{r}; \mathbf{R}_0), \tag{5.134c}$$

where for this system we write $W_n(\mathbf{R}) = E_n^{(0)} + \alpha^2 W_n^{(2)} + \cdots$. A comparison of Eqs. (129) and (134) shows that they differ only in Eq. (134c), i.e., in the

---

† Although the potential $V(\mathbf{r}, \mathbf{R})$ contains terms that depend on the nuclear coordinates $\mathbf{R}$, the effect of these is contained in $\psi_n(r; \mathbf{R})$ and $E_n(\mathbf{R})$, where these coordinates are just parameters.

second order of perturbation theory. We multiply Eqs. (134b) and (134c) by $\chi_n^{(1)}(\mathbf{R})$ and $\chi_n^{(0)}(\mathbf{R})$, respectively, and inserting them in Eq. (133), we find that

$$(H_0^{(0)} - E_n^{(0)})[\Psi_n^{(2)}(\mathbf{r}; \mathbf{R}) - \chi_n^0(\mathbf{R})\psi_n^{(2)}(\mathbf{r}; \mathbf{R}) - \chi_n^{(1)}(\mathbf{R})\psi_n^{(1)}(\mathbf{r}; \mathbf{R})]$$

$$= -(H_N^{(2)} + W_n^{(2)} - E_n^{(2)})\chi_n^{(0)}(\mathbf{R})\psi_n^{(0)}(\mathbf{r}; \mathbf{R}_0). \qquad (5.135)$$

This equation has a nontrivial solution only if its right-hand side is orthogonal to the solution of the homogeneous equation (134a), i.e., if

$$\int \psi_n^{(0)}(\mathbf{r}; \mathbf{R}_0)(H_N^{(2)} + W_n^{(2)} - E_n^{(2)})\chi_n^0(\mathbf{R})\psi_n^{(0)}(\mathbf{r}; \mathbf{R}_0) \, dv = 0. \quad (5.136)$$

Since the operator in the integrand does not depend on the electron coordinates $\mathbf{r}$, Eq. (136) will be satisfied if and only if

$$(H_N + \alpha^2 W_n^{(2)} - \alpha^2 E_n^{(2)})\chi_n^{(0)}(\mathbf{R}) = 0. \qquad (5.137)$$

Equations (134a) and (137) determine the two factors in the zeroth-order wave function, $\Psi_n^{(0)}$. written in the form of Eq. (131). It follows from these equations that the energy up to second order corresponding to this wave function is equal to the sum of the energy of the electrons moving around a system of stationary nuclei and that of the nuclei moving in an effective potential $\alpha^2 W_n^{(2)}$.

The results that we have just obtained have a simple physical interpretation. The velocities of the electrons and of the nuclei are very different because of the great difference in their masses. As a result, the electrons move so rapidly that they see the nuclei as practically unmovable point charges. The slow moving nuclei, on the other hand, experience only the average motion of the electron cloud that surrounds them, and this is why only the average energy of the electrons appears in Eq. (137). This approximation is known as the adiabatic approximation because the electrons adjust adiabatically to each new configuration of the nuclei. In higher orders of perturbation theory, however, the system's wave function can no longer be expressed as the product of a function of the nuclear coordinates $\mathbf{R}$ and a function of the electron coordinates $\mathbf{r}$ in which $\mathbf{R}$ enters solely as a parameter. Rather, it must be expressed as the sum of such products, as in Eq. (132).

Since the terms $E_n^{(2)}$ and $W_n^{(2)}$ are associated with the second powers of $(\mathbf{R} - \mathbf{R}_0)$, Eq. (137) describes the motion of the nuclei in the harmonic approximation. As we saw in Section 1.4, this motion can be treated by introducing normal coordinates, and these permit, for instance, a simple description of the mechanical and statistical properties of crystal lattices.

Let us now consider an equation such as (134a) for the electronic wave function $\psi_n^{(0)}(\mathbf{r}; \mathbf{R}_0)$. In view of Eqs. (122) and (123), this equation can be rewritten in the form

$$\left[\sum_i \frac{-\hbar^2}{2m} \nabla_i^2 - \sum_{i,a} \frac{Z_a e^2}{|\mathbf{r}_i - \mathbf{R}_a|} + \sum_{i \neq j} \frac{e^2}{|\mathbf{r}_i - \mathbf{r}_j|}\right] \psi_n(\mathbf{r}; \mathbf{R}) = E_n(\mathbf{R})\psi_n(\mathbf{r}; \mathbf{R}),$$

(5.138)

where we have incorporated into $E_n(\mathbf{R})$ the term $\sum_{a \neq b} Z_a Z_b e^2/|\mathbf{R}_a - \mathbf{R}_b|$ of $V(\mathbf{r}, \mathbf{R})$. The first two terms on the left-hand side of this equation describe the electron states in the field of the nuclei, while the third term relates to the electron–electron interaction.

For very simple systems, Eq. (138) can be solved explicitly for different fixed positions of the nuclei, leading to an exact expression for the electron's energy in place of a perturbation expansion $E_n(\mathbf{R}) = E_n^{(0)} + \alpha^2 W_n^{(2)} + \cdots$. For instance, the hydrogen molecular ion H_2^+ contains just one electron and two nuclei, so that $E_n(\mathbf{R})$ is a function only of the distance R between the nuclei, while there is no electron–electron interaction. Calculations lead, for the energy $E_0(R)$ of the ground state, to the graph shown in Fig. 4, and it is clear that the equilibrium internuclear distance R_0 is just that at which $E_0(R)$ has a minimum. The ground states of a few other molecules that contain only a very small number of particles can be found in a similar way. The results of such calculations are found to be in excellent agreement with experimental measurements of the internuclear distances and electron energies. However, for larger molecules, the calculations become much more cumbersome and the agreement between them and the experimental results is much poorer.

For macroscopic systems containing large numbers of particles, it is quite impractical to solve Eqs. (134) or (138) without making additional approximations. A crude model for the electron states in a crystalline solid,

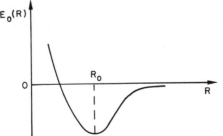

Fig. 4. The lowest energy level $E_0(R)$ of the H_2^+ ion as a function of the internuclear distance R.

for instance, involves an assumption that the nuclei are arranged on the sites of a perfect crystal lattice, and the replacement of the electron–electron interaction by an effective potential experienced by each individual electron. Surprisingly enough, even such a simple model can predict quite well the energy spectrum for electrons in a crystal.

Degenerate Systems with Two Perturbations

The adiabatic approximation is in fact an example of a system containing two perturbations, namely, the kinetic energy of the ions and the effect of their motion on the electronic wave functions. Since the ground state was not degenerate, it was possible to treat these two perturbations simultaneously. However, difficulties in the choice of the relevant small parameter can arise for systems whose unperturbed operator H_0 has degenerate eigenstates. If there is only a single perturbation, this determines which combinations of the degenerate eigenfunctions of H_0 are to be used as a first approximation to the eigenfunctions of the perturbed system. The subsequent development of perturbation theory for the system then involves a lot of complicated algebra, but no difficulties of principle. These will only arise if there are two or more perturbations, each of which on its own would lead to a different first approximation for the perturbed system's eigenfunctions.

As an example of such a system, we will consider the nature of the electron states in disordered materials such as amorphous or liquid metals and amorphous semiconductors. In a perfect crystal, the symmetry requirements discussed in Section 3.3 ensure that all the eigenfunctions extend over the whole crystal, with just a change of phase $\mathbf{k} \cdot \mathbf{n}$ on passing from one lattice site to another distance $\mathbf{n}$ away [see Eq. (3.112)]. In amorphous materials, however, there is no translational symmetry or long-range order, so that these requirements no longer exist. The question then arises whether the eigenfunctions, i.e., the wave functions of the electrons, still extend over a long distance, or whether they are now localized. Only if the states are extended can we expect a high electron mobility and electrical conductivity in disordered systems, a feature that is of great practical importance. Experimental results show that in liquid metals the electron states are extended, but that in amorphous semiconductors they are sometimes extended and sometimes localized.

In the simplest possible realistic model of an amorphous system, we consider a system of identical atoms situated at fixed points $\mathbf{R}_j$, and assume that the electronic wave functions ψ can be expressed as linear combinations of a single function $\phi_j(\mathbf{r}) \equiv \phi(\mathbf{r} - \mathbf{R}_j)$ per atom. These functions are assumed, for convenience, to be normalized and mutually orthogonal. Instead of defining a Hamiltonian H explicitly, it is sufficient if we define its matrix elements between the different functions ϕ_j. Thus, we write

$$H = H_0 + H_1 + H_2, \qquad (5.139)$$

where in the simplest case

$$(H_0)_{jk} = \varepsilon_0 \delta_{j,k}, \qquad (5.140)$$

$$(H_1)_{jk} = \varepsilon_j \delta_{j,k}, \qquad (5.141)$$

$$(H_2)_{jk} = \begin{cases} V & \text{if atoms } j \text{ and } k \text{ are nearest neighbors} \\ 0 & \text{otherwise.} \end{cases} \qquad (5.142)$$

The first term H_0 represents the average energy of an individual atom. H_1 contains the random part ε_j of the energy H_{jj} of ϕ_j, which we call the site energy, and which arises because the atoms are not all in exactly the same environment. H_2 expresses the probability of an electron's tunneling from one atom to another, and the assumption that V is a constant is made so as to simplify the analysis. We consider a situation in which ε_j is a random variable uniformly distributed between $-\frac{1}{2}W$ and $\frac{1}{2}W$.

For this system, the zeroth-order Hamiltonian H_0 is completely degenerate, with all its eigenvalues equal to ε_0. In the absence of H_2, i.e., if $V = 0$, the term H_1 would remove this degeneracy and the eigenfunctions of H would be the localized functions ϕ_j, with eigenvalues $\varepsilon_0 + \varepsilon_j$. This is effectively the situation in a gas, for instance, where the atoms are too far apart for V to be appreciable. If H_1 is missing, on the other hand, i.e., if $W = 0$, the Hamiltonian $H_0 + H_2$ is similar to that for a crystal, so that the eigenfunctions extend over the whole system. For this case, it can readily be shown that the eigenvalues are distributed continuously throughout a band centered on ε_0 and of width approximately $2ZV$, where Z is the average number of nearest neighbors (or coordination number) of each atom. We expect that if $W \gg 2ZV$, then H_2 should be treated as a perturbation to $H_0 + H_1$, so that V (or V/W) is the appropriate small parameter. If, on the other hand, $2ZV \gg W$, the randomness in the energies ε_j should have only a small effect, so that H_1 can be treated as a small perturbation and W/V is the relevant small parameter. Difficulties arise when W and $2ZV$ are of the same order of magnitude, so that H_1 and H_2 would each independently produce the same sort of spread in the eigenvalues of H. Although this problem has been extensively studied, no exact solutions are known for a realistic geometry of a three-dimensional system. However, a very interesting qualitative approach, which involves a novel method of applying perturbation theory, has recently been proposed, and we will now present and discuss this.

Flexible Choice of the Perturbation

Instead of treating the eigenvalue E of H, i.e., the energy, as an unknown, we start by fixing its value and looking for the corresponding eigenfunctions, if there are any. This approach is, of course, ideally suited to Brillouin–Wigner perturbation theory, which, as we saw previously, is much more suitable for the solution of degenerate problems than is Rayleigh–Schrödinger theory. If E lies well within the range of the site energies $\varepsilon_0 + \varepsilon_j$, we expect the eigenfunctions to have their largest amplitudes on sites whose energy is close to E. Thus, our first step is to consider only the matrix elements of H between sites whose energies lie in the range $E \pm yV$, where y is a constant of order unity. We treat these elements of H as our zeroth-order Hamiltonian, which we now denote by H_E, and all the other elements as a perturbation. In other words, the choice of what is regarded as a perturbation depends on E, instead of being fixed as in ordinary perturbation theory. The set of atoms involved in H_E constitute a fraction $(2yV/W)$ of the total. If this number is large enough for a continuous path, going from one atom to a nearest neighbor at each step, to be constructed throughout the system from this set, we expect the eigenfunctions to be extended (since H_2 allows the rapid passage of an electron from one site to another in this set). The critical values of $(2yV/W)$ for this to happen is found by percolation theory to be close to $2/Z$.

The most important effect of the sites not included in H_E will be to provide a coupling between the atoms in our original set that are not nearest neighbors. The site energy of an atom not in our original set will typically differ from that of one inside the set by an amount of order $W/4$, so that in second-order perturbation theory next nearest neighbor atoms included in H_E will be coupled by an effective matrix element $V^2/(W/4) = 4V^2/W$. We therefore define a new zeroth-order Hamiltonian, denoted by $H_E^{(\mathrm{I})}$, which includes H_E plus the matrix elements associated with all atoms whose site energy is within $4V^2/W$ of those in H_E. In other words, $H_E^{(\mathrm{I})}$ includes all states with energy in the range $E \pm yV \pm 4V^2/W$. If a path throughout the system, going from one atom to a nearest neighbor or a next nearest neighbor can be constructed with these atoms, we once again expect to obtain extended eigenfunctions. Otherwise, we must consider also the effect of second nearest neighbors, define a new zeroth-order Hamiltonian $H_E^{(\mathrm{II})}$, and so on. If none of these Hamiltonians leads to extended states, we can expect that any eigenstates of energy E are localized. For E outside the main range of $(\varepsilon_0 + \varepsilon_j)$, the appropriate modification of the above procedure can readily be derived.

Whether or not this method can be used for quantitative calculations on amorphous solids, it provides a good example of how the zeroth-order

Hamiltonian appropriate to a problem need not be fixed in advance, but can be chosen self-consistently for the eigenvalue of interest.

5.4   Difficulties in the Use of the Small Parameter

Throughout this chapter, we have emphasized the expansion of the solution to a problem involving a small parameter as a power series in this parameter, in which the term of zeroth order is the solution of the unperturbed problem. However, difficulties sometimes arise in the use of such a procedure. One example of such difficulties occurs if the small parameter multiplies the highest derivative in a differential equation. In this situation, which we discuss below, the expansion method is usually not directly applicable.

A rather different case in which its application is not straightforward is when the unperturbed system has a set of degenerate eigenfunctions. In that case, as we saw at the end of the last section, it is the perturbation that determines the appropriate zeroth-order approximation to the solution. This method is closely related to the use of a renormalized energy, as we saw in Eq. (53). Such a renormalization procedure can also be essential in other cases, such as for the oscillations of an anharmonic oscillator considered in Section 5.1.

Another sort of problem, to which we denote the last part of this section, is one whose zeroth-order solution is not known exactly, but where a knowledge of some of its properties enables us to derive useful results. For instance, whatever the exact form of the eigenfunctions of the Hamiltonian of a perfect crystal, we know from Section 3.3 that they must have a definite translational symmetry. This fact, as we will see, is sufficient to enable us to obtain useful results for such diverse problems as impurity states in semiconductors and the interaction between nuclear magnetic moments in metals.

A Small Parameter Multiplying the Highest Derivative

Apart from the rather artificial example of an oscillator whose mass tends to zero which we discussed in Section 5.1, we have met in this book three different problems in which the highest derivative of the unknown function with respect to some variable is multiplied by a small parameter. In Section 1.5, when we discussed space-charge-limited currents, the diffusion term is the one with the highest derivative, and we ignored this completely in order to obtain a problem that was mathematically tractable. For the problem from hydrodynamics in Section 1.6, on the other hand, where the term associated with the fluid's viscosity involves the derivative of higher order, we assumed that this term was the dominant one in the boundary layer and could be ignored elsewhere. The third example, considered briefly in Section 2.4, was that of heat conduction in a thin plate, of thickness L, length M, and width N. Here, the small parameters $(L/M)^2$ and $(L/N)^2$ in Eq. (2.122) multiply the terms $\partial^2 F/\partial v^2$ and $\partial^2 F/\partial w^2$ associated with the flow of heat and temperature changes in the plane of the plate. We ignored these terms, thereby removing all dependence of the solution on v and w, even though the resulting solution does not satisfy the boundary conditions on the edges of the plate.

The reasoning behind and justification for these approaches is different in each case. For the flow of fluid past a solid surface, the boundary condition that the tangential component of the fluid's velocity at the surface should vanish [Eq. (1.29)] has a clear physical meaning, and cannot readily be ignored or replaced by a boundary condition at some other point. Therefore, our solution must satisfy this condition at the surface, and we use the boundary-layer theory described in Section 1.6 to achieve this. For space-charge-limited currents, on the other hand, even if we were to introduce a boundary layer in which the diffusion term was the dominant one, the equations for the carrier density in each region could still not readily be solved. In addition, and much more importantly, the boundary condition at the electrode $x = 0$ from which electrons enter the insulator does not usually involve the carrier density as a separate variable. Instead of a condition being imposed at $x = 0$, a corresponding condition can be imposed at a point $x = a$, close to the origin, beyond which the diffusion term is in fact negligible. This point corresponds to that where the approximate and exact solutions of the damped harmonic oscillator shown in Fig. 3 coincide. We do not lose much physical information by ignoring the region between $x = 0$ and $x = a$, while if we ignored the boundary layer in the fluid flow we would lose all information about the resistance to the solid's motion through the fluid.

In the problem of the thin plate, the justification for ignoring the dependence of F on v and w, or equivalently of the temperature T on the coordinates y and z, is that this dependence is not important. Physically, this is so because heat is expected to diffuse to or from the edges of the plate at approximately the same rate in all directions. As a result, only a very small fraction of the volume of the plate, very close to its edges, will change its temperature more rapidly as a result of the diffusion of heat across the edges rather than across the main surface of the plate. Mathematically, this result follows from the exact solution of the relevant heat diffusion problem, Eqs. (2.114) and (2.115).

The Effective Mass Approximation

We now turn to problems in which use is made only of the symmetry of the eigenfunction of the unperturbed system, and not of the explicit function, which is in most cases not well known. The two problems that we consider are associated with electrons in crystals. In accordance with our discussion in the last section, following the derivation of the adiabatic approximation, we consider an unperturbed one-electron Hamiltonian H_0 of the form

$$H_0 = (-\hbar^2/2m)\,\nabla^2 + V_p(\mathbf{r}), \tag{5.143}$$

where $V_p(\mathbf{r})$ is a potential having the periodicity of the crystal lattice. As we saw in Section 3.3, the eigenfunctions of such a Hamiltonian are of the form $\psi_k(\mathbf{r})$, where

$$\psi_k(\mathbf{r} + \mathbf{n}) = \exp(i\mathbf{k}\cdot\mathbf{n})\psi_k(\mathbf{r}). \tag{5.144}$$

If the values of $\mathbf{n}$ are restricted to the first Brillouin zone, each set of functions ψ_k defines a set of states or band s, whose wave functions we will denote by $\psi_{k,s}(\mathbf{r})$.

The first problem that we consider is the effect on the eigenfunctions and eigenvalues of an additional, fairly localized, potential $w(\mathbf{r})$, such as that produced by an impurity atom. Not surprisingly, it proves more convenient to discuss these effects in terms of localized functions rather than in terms of the extended eigenfunctions $\psi_{k,s}$. We make use of the periodicity of $\psi_{k,s}$, as described by Eq. (144), to define Wannier functions $a_s(\mathbf{r})$ by means of the Fourier transform pair

$$\psi_{k,s}(r) = \frac{1}{\sqrt{N}}\sum_{\mathbf{n}} \exp(i\mathbf{k}\cdot\mathbf{n})a_s(\mathbf{r} - \mathbf{n}), \tag{5.145}$$

and

$$a_s(\mathbf{r} - \mathbf{n}) = \frac{1}{\sqrt{N}}\sum_{k} \exp(-i\mathbf{k}\cdot\mathbf{n})\psi_{k,s}(\mathbf{r}), \tag{5.146}$$

where N is the number of lattice sites $\mathbf{n}$. It is impossible to treat the additional potential $w(r)$ by means of standard perturbation theory, since we do not know the solution of the unperturbed problem. Instead, we look for eigenfunctions ψ of $H_0 + w$ of the form

$$\psi(\mathbf{r}) = \frac{1}{\sqrt{N}}\sum_{s,\mathbf{n}} f_s(\mathbf{n})a_s(\mathbf{r} - \mathbf{n}). \tag{5.147}$$

We then obtain a set of difference equations for the coefficients $f_s(\mathbf{n})$, which can be replaced by differential equations for functions $f_s(\mathbf{r})$ of the continuous variable $\mathbf{r}$ provided that $w(\mathbf{r})$ is a smooth function of $\mathbf{r}$, i.e. does not change much from one lattice site to the next. If, in addition, $w(\mathbf{r})$ is not too strong, it is sufficient to consider each band of states separately, and we can omit the summation over s. In the absence of $w(\mathbf{r})$, we would find, on comparing Eqs. (145) and (147), that f_s reduces to $f_s^0(\mathbf{r}) = \exp(i\mathbf{k} \cdot \mathbf{r})$. Let us define, for band s, an effective mass m_s^* (at least for $\mathbf{k}$ close to the origin) by the Taylor series for $E_s(\mathbf{k})$ near $\mathbf{k} = 0$,†

$$E_s(\mathbf{k}) = E_s(0) + (\hbar^2/2m_s^*)k^2. \tag{5.148}$$

We see that in the absence of $w(\mathbf{r})$ the function f_s^0 satisfies the equation

$$(-\hbar^2/2m_s^*)\,\nabla^2 f_s^0(\mathbf{r}) = [E_s(\mathbf{k}) - E_s(0)]f_s^0(\mathbf{r}). \tag{5.149}$$

As is shown in many textbooks on solid state physics, in the presence of the potential $w(\mathbf{r})$, $f_s(\mathbf{r})$ satisfies the natural generalization of this equation

$$(-\hbar^2/2m_s^*)\,\nabla^2 f_s(\mathbf{r}) + w(\mathbf{r})f_s(\mathbf{r}) = [E_s - E_s(0)]f_s(\mathbf{r}). \tag{5.150}$$

Thus, a knowledge of the eigenvalues $E_s(\mathbf{k})$ and hence of m_s^* is sufficient to determine, in this case, the eigenvalues E_s and the envelope function $f_s(\mathbf{r})$ for the perturbed problem. In many cases, it is very difficult to calculate m_s^*, and this effective mass is treated as a parameter to be determined from experiments.

*Magnetic Interactions of Nuclei Through Conduction Electrons

Another problem in which only the form of the zeroth-order eigenfunctions is used is that of magnetic interactions in metals. The interaction between the nuclear magnetic moments, which can be observed from atomic spectra, is believed to take place in many metals by an interaction between the conduction electrons and the magnetic moments of the different nuclei. If we consider a simple crystal in which the nucleus at lattice site $\mathbf{n}$ has magnetic moment $\mathbf{I_n}$, this interaction causes the electrons to experience a potential of the form

$$v(\mathbf{r}) = A \sum_{\mathbf{n}} (\mathbf{I_n} \cdot s)\, \delta(\mathbf{r} - \mathbf{n}), \tag{5.151}$$

where s is the spin of the electron. We now assume that the direction of an electron's spin can be changed much more easily than those of the nuclear spins, so that we can treat the two sets of spins by the Born–Oppenheimer approximation described in Section 5.3. The zeroth-order Hamiltonian is that appropriate to the conduction electrons, and we do not know its eigen-

---

† The term linear in $\mathbf{k}$ in the Taylor series vanishes at $\mathbf{k} = 0$ for reasons of symmetry.

functions. In spite of this, we can derive the effect of $v(\mathbf{r})$ on the nuclear spins. In accordance with Eq. (137), the nuclear spins $\mathbf{I_n}$ will experience an interaction potential $W^{(2)}$ which is just the sum, over all the conduction electrons, of the changes in the electrons' energy to second order in perturbation theory as a result of the interaction potential $v(\mathbf{r})$. We denote by $v_{\mathbf{k},\mathbf{k}'}$ the matrix element of $v(\mathbf{r})$ between the electron states of wave vector $\mathbf{k}$ and $\mathbf{k}'$. Then, at $0°K$, when all the states below the Fermi level, i.e., with $|k| < k_F$, are occupied, the only processes that can contribute to the energy in second-order perturbation theory are those in which electrons are excited virtually to a state with $|k| > k_F$ and return. Thus

$$W^{(2)} = \sum_{|k|<k_F} \sum_{|k'|>k_F} \frac{|v_{\mathbf{k},\mathbf{k}'}|^2}{E(\mathbf{k}) - E(\mathbf{k}')}. \tag{5.152}$$

For a crystal of unit volume, if the sums over $\mathbf{k}$ and $\mathbf{k}'$ are replaced by integrals, this expression becomes

$$W^{(2)} = \left(\frac{1}{2\pi}\right)^6 \int_{|k|<k_F} d\mathbf{k} \int_{|k'|>k_F} d\mathbf{k}' \frac{|v_{\mathbf{k},\mathbf{k}'}|^2}{E(\mathbf{k}) - E(\mathbf{k}')}. \tag{5.153}$$

In order to evaluate $v_{\mathbf{k},\mathbf{k}'}$, we make use of the fact that the electronic wave functions in the absence of $v(\mathbf{r})$ are of the form given by Eq. (144). Hence, if we write

$$\psi_{\mathbf{k}}(\mathbf{r}) = \exp(i\mathbf{k}\cdot\mathbf{r})u_{\mathbf{k}}(\mathbf{r}), \tag{5.154}$$

the functions $u_{\mathbf{k}}(\mathbf{r})$ will have the periodicity of the crystal lattice, i.e.,

$$u_{\mathbf{k}}(\mathbf{r} + \mathbf{n}) = u_{\mathbf{k}}(\mathbf{r}), \tag{5.155}$$

as we saw in Chapter 3. Thus, in view of the delta-functions in $v(\mathbf{r})$,

$$\begin{aligned} v_{\mathbf{k},\mathbf{k}'} &= A\sum_{\mathbf{n}} (\mathbf{s}\cdot\mathbf{I_n})\psi_{\mathbf{k}}^*(\mathbf{n})\psi_{\mathbf{k}'}(\mathbf{n}) \\ &= A\sum_{\mathbf{n}} (\mathbf{s}\cdot\mathbf{I_n})u_{\mathbf{k}}^*(0)u_{\mathbf{k}'}(0) \exp(i(\mathbf{k} - \mathbf{k}')\cdot\mathbf{n}). \end{aligned} \tag{5.156}$$

Hence,

$$W^{(2)} = A^2(2\pi)^{-6} \sum_{\mathbf{m}}\sum_{\mathbf{n}}(\mathbf{s}\cdot\mathbf{I_m})(\mathbf{s}\cdot\mathbf{I_n}) \iint d\mathbf{k}\, d\mathbf{k}'\, |u_{\mathbf{k}}^*(0)u_{\mathbf{k}'}(0)|^2$$

$$\times \frac{\exp[i(\mathbf{k} - \mathbf{k}')\cdot(\mathbf{n} - \mathbf{m})]}{E(\mathbf{k}) - E(\mathbf{k}')}, \tag{5.157}$$

where the limits of integration are as in Eq. (152). Hence $W^{(2)}$ can be expressed as the sum of terms $H_{\mathbf{m},\mathbf{n}}$ that correspond to an interaction between the nuclear spins at sites $\mathbf{m}$ and $\mathbf{n}$. Since the average over the electron spin $\mathbf{s}$ of $(\mathbf{s} \cdot \mathbf{I_m})(\mathbf{s} \cdot \mathbf{I_n})$ is proportional to $\mathbf{I_m} \cdot \mathbf{I_n}$, and $H_{\mathbf{m,n}}$ otherwise involves only $\mathbf{m} - \mathbf{n}$, we can write

$$W^{(2)} = \sum_{\mathbf{m,n}} H_{\mathbf{m,n}} = \sum_{\mathbf{m}} \sum_{\mathbf{n}} J(\mathbf{m} - \mathbf{n}) \mathbf{I_m} \cdot \mathbf{I_n}. \tag{5.158}$$

In order to estimate the integral $J(\mathbf{m} - \mathbf{n})$, we assume that, at least for all important values of $\mathbf{k}$ and $\mathbf{k'}$, $|u_{\mathbf{k}}^*(0)u_{\mathbf{k'}}(0)|$ is independent of $\mathbf{k}$ and $\mathbf{k'}$ and that $E(\mathbf{k})$ is given by the effective mass approximation, Eq. (148). In that case, the integral in Eq. (157) can readily be evaluated, and it is found that, if we write $|\mathbf{m} - \mathbf{n}| = R_{mn}$, then

$$J(\mathbf{m} - \mathbf{n}) \propto R_{mn}^{-4}[2k_F R_{mn} \cos(2k_F R_{mn}) - \sin(2_F R_{mn})]. \tag{5.159}$$

Thus, $J(\mathbf{m} - \mathbf{n})$ decreases rapidly as R_{mn} increases. It is negative, so that parallel spins are energetically more favorable, if $R_{mn} < \pi/4k_F$, while for larger values of R_{mn}, its sign oscillates. We note that we did not require the explicit form of $\psi_{\mathbf{k}}(\mathbf{r})$ or of $E(\mathbf{k})$ to obtain Eq. (158), while even the approximate value of $J(\mathbf{m} - \mathbf{n})$ only requires a knowledge of the effective mass.

Problems

5.1 An interesting and simple example of the application of perturbation theory to a problem in classical mechanics is provided by the following problem. Consider a pendulum consisting of a flexible cord of mass m and length l, at one end of which there is a heavy bob of mass $M \gg m$, while the other end is fixed. We wish to calculate the effect of the finite value of m on the pendulum's period.

(a) Assuming that the pendulum executes small oscillations with angular frequency ω, show that the displacement in the horizontal direction $y(x)$ of the cord at a point distance x from the bob satisfies the equation $(1 + \lambda x/l)y'' + (\lambda/l)y' + (\lambda\omega^2/l^2\omega_0^2)y = 0$, where $\omega_0 = (g/l)^{1/2}$ is the period of a pendulum of length l with a massless string and $\lambda = (m/M)$ is the system's small parameter. Show also that the boundary conditions for this problem are $y(l) = 0$ and $y'(0) = -(\omega^2/\omega_0^2)y(0)/l$.

(b) Solve the above problem to first order in λ, by writing $y = y_0 + \lambda y_1 + \cdots$, $\omega^2 = \omega_0^2 + \lambda\omega_1^2 \cdots$. Hence, show that the effect of the mass of the cord, to first order in (m/M), is to reduce the pendulum's period by a factor of $1 - \frac{1}{12}(m/M)$. See H. L. Armstrong, *Amer. J. Phys.* **44**, 565 (1976); S. T. Epstein and M. G. Olson, *Amer. J. Phys.* **45**, 671 (1977).

5.2   The equation of motion of an electron of mass m bound to the nucleus at the origin by a one-dimensional restoring force $-m\omega_0^2 x$, and subject to radiation damping, can be expressed in the form (as show in the reference noted below)

$$\ddot{x}(t) + \omega_0^2 \sum_{n=0}^{\infty} \tau^n \, d^n x/dt^n = 0,$$

where $\tau = \frac{2}{3}(e^2/mc^3)$ in Gaussian units, and has the dimensions of time.

(a)   By expressing the equation in dimensionless form, show that the natural small parameter of the system is $\lambda = \omega_0 \tau$.

(b)   Solve this system of equations up to second order in λ

(i)   by expanding x as a power series in λ;

(ii)   by assuming that $x = x_0 \exp(-\alpha t)$, and expanding α as a power series in λ.

Compare the results, and from (ii) find the frequency shift and rate of decay produced by the radiation damping. See P. Chand, *Amer. J. Phys.* **39**, 338 (1971).

5.3   Show that the choice of $y_0 = au_0$, where a is some arbitrary constant, in place of $y_0 = u_0$ in Rayleigh–Schrödinger perturbation theory, does not affect the value of the terms ε_j that appear in the expansion of the energy, Eq. (32). See S. T. Epstein, *Amer. J. Phys.* **36**, 165 (1968).

5.4   (a)   Find by Rayleigh–Schrödinger perturbation theory, to second order in λ, the eigenvalues of the following problem, which is just Schrödinger's equation for a particle enclosed in a one-dimensional box and subject to an attractive δ-function potential:

$$(-\hbar^2/2m) \, d^2\psi/dx^2 - \lambda\delta(x)\psi(x) = E\psi(x), \qquad \psi(L) = \psi(-L) = 0.$$

(b)   Express the above equation and its boundary conditions in dimensionless form, and hence show that the correct small parameter in terms of which the energy should be expanded is not λ but rather $(2m/\hbar^2)\lambda L$. See S. T. Epstein, *Amer. J. Phys.* **28**, 495 (1960).

5.5   For the Sturm–Liouville operator $H_0 = (d/dx)[p(x) \, d/dx] + q(x)$, it is possible to construct an explicit expression for the Green's function $G_0(x, x', E)$ appropriate to the problem

$$(H_0 - E)y = 0,$$

with homogeneous boundary conditions $a_1 y(a) + a_2 y'(a) = 0$, $b_1 y(b) + b_2 y'(b) = 0$, as follows.

(a)   Let $u(x)$ be a solution of $(H_0 - E)y = 0$ satisfying the boundary condition at $x = a$, and $v(x)$ a solution of this equation satisfying the boundary condition at $x = b$. Show that if E is not an eigenvalue of the problem, $u(x)$ and $v(x)$ will be linearly independent, and that in such a case

$$G_0(x, x', E) = \begin{cases} -(1/A)u(x)v(x'), & x < x', \\ -(1/A)u(x')v(x), & x > x' \end{cases}$$

where $A = p(x)[u(x)v(x') - u(x')v(x)]$, satisfies Eq. (94) and so is the problem's Green's function.

(b) For the operator $H_0 = (-d^2/dx^2)$, show that the boundary conditions that the solution represents an outgoing wave at $x = \pm\infty$ can be written in the prescribed form, and that the above formula leads to the Green's function

$$G_0(x, x', E) = (i/2k) \exp(ik|x - x'|), \qquad \text{where} \quad k = \sqrt{E}.$$

See G. Arfken, "Mathematical Methods for Physicists," Section 16.5, Academic Press, New York, 1966.

5.6 Use the result of Problem 5.5 to derive an integral equation, analogous to Eq. (108), for the scattering of an incoming free electron with wave function $\psi(x) = \exp(ikx)$ by a δ-function potential at the origin, $V(x) = \lambda\delta(x)$. Solve this equation iteratively to all orders in λ to obtain the scattering matrix for this system, and compare your result with that of Problem 4.2. See I. R. Lapidus, *Amer. J. Phys.* **37**, 1064 (1969).

5.7 Apply the first Born approximation, Eq. (108), to the problem of the scattering of an electron from the three-dimensional square well potential of Eq. (4.56), and hence obtain the result given in Eq. (4.61).

*5.8 In Eq. (98), we presented an expression for the Green's function $G_0(\mathbf{r}, \mathbf{r}', E)$ of the operator H_0 in terms of its eigenfunctions u_k and eigenvalues E_k. In this representation, we can write for the trace of the Green's function

$$\text{Tr } G_0(\mathbf{r}, \mathbf{r}', E) \equiv \sum_k \iint u_k^*(\mathbf{r})u_k(\mathbf{r}')G_0(\mathbf{r}, \mathbf{r}', E) \, dv \, dv'.$$

(a) Show that $\text{Tr } G_0(\mathbf{r}, \mathbf{r}', E) = \sum_k 1/(E - E_k)$.

(b) For a system with a continuum of eigenvalues, such that the density of eigenvalues lying between E and $E + dE$ is $g(E) \, dE$, the natural extension of the result of (a) is

$$G(\mathbf{r}, \mathbf{r}' E) = \int_{-\infty}^{\infty} \frac{g(\eta)}{(E - \eta)} \, d\eta.$$

Using this equation show that

$$g(E) = -\frac{1}{\pi} \lim_{\delta \to 0} \text{Im Tr } G(\mathbf{r}, \mathbf{r}', E + i\delta).$$

The practical importance of this result arises from the fact that there exist a number of ways of deriving the Green's function, such as that described in Problem 5.4 or the solution of Dyson's equation, that do not involve first finding the eigenvalues and eigenfunctions of the operator with which it is associated.

*5.9 Apply the Born–Oppenheimer approximation described in Section 5.3 to the hydrogen atom (without separating the motion of the center of mass from the rest of the problem, of course). The simplicity of this problem makes it practicable to go one stage beyond the adiabatic approximation, and to express Ψ for the electronic ground state in the form $\Psi = \Sigma_n z_n(\mathbf{R})\psi_n(\mathbf{r}; \mathbf{R})$, where the functions ψ_n are the eigenfunctions of H_0, rather than just use the first term of such a sum, as in Eq. (131). Do this, and compare the results with the exact ones obtained by solving the problem in center-of-mass coordinates. See V. K. Deshpande and J. Mahanty, *Amer. J. Phys.* **37**, 823 (1969).

Chapter 6 | Epilogue—Example of the Application of the Above Methods to a Problem in Nonlinear Optics

6.1 Preface

Starting with a physical problem, we have seen in the preceding chapters how to construct a model that is simpler to analyze (Chapter 1), determine from dimensional analysis how the solution depends on the relevant physical parameters (Chapter 2), use symmetry (Chapter 3) and general analytical considerations (Chapter 4) to establish various properties of the solution, and finally how to use the method of the small parameter (Chapter 5) to find its order of magnitude. In this chapter, we demonstrate the application of these methods to a problem of practical interest and importance.

In principle, the solution of any physical problem can be made easier by the application of these methods, and we hope that the reader will apply them to the problems that confront him. To illustrate how this process can be used in practice, we have chosen (almost at random) to examine the problem of second harmonic generation in optics. This consists of the use of the non-linear optical properties of a crystal to convert a beam of light of a given frequency to one of double that frequency.

In order to study this problem, we start by constructing and examining a model of a system possessing nonlinear optical properties. This model is useful because it leads naturally to ideas of how frequency doubling and similar processes can occur, while it is simple enough to be solved exactly, as well as by means of dimensional analysis. We next use response functions and the consequences of causality to provide a more formal and rigorous definition of nonlinear susceptibilities, which we insert into an expression for the system's free energy in order to derive its intrinsic symmetry properties. For a specific crystal, in this case KDP (potassium dihydrogen phosphate), we then use the system's rotational symmetry to find which elements are zero and which are equal to each other in the nonlinear susceptibility tensor that is relevant to second harmonic generation.

After deriving all these general properties, we turn to a calculation of the rate at which second harmonic generation will take place. To do this, we first derive the equations that the electric fields of the light waves must satisfy. Since these equations are nonlinear and cannot be solved exactly, we have to prepare the problem for the application of perturbation theory, after which we apply it to a specific simple case to derive explicit formulas. Throughout this chapter, we use Gaussian rather than S.I. electrical units, since these are at present the most widely used in books and articles on nonlinear optics.

6.2 Model System

Analysis of the Model

As a simple model for a medium exhibiting nonlinear optical effects, we consider a set of identical, noninteracting particles (atoms, molecules, or even unit cells of crystals) in each of which an electric field induces a dipole moment. In order to obtain a model that can be analyzed easily, we examine a classical system in which each particle can be regarded as consisting of a charge q of mass m which is bound to a fixed center by a force that depends only on its position $\mathbf{r}$ relative to that center. As we saw in Section 4.1, this simple model leads to a useful description of a medium's dielectric constant, and hence of its linear optical properties, if the force on the charge is proportional to its distance from the center, so that it behaves like a harmonic oscillator. The natural extension of this model to obtain nonlinear effects is to consider an anharmonic oscillator. Since tensor effects are of considerable importance, we consider a three-dimensional oscillator rather than a linear one. Thus, we write the equation of motion of a typical particle in the form

$$\ddot{x}_j + a\dot{x}_j + \sum_l K_{jl}x_l + \sum_l \sum_n c'_{jln}x_l x_l x_n = \frac{q}{m}E_j \exp(-i\omega t), \qquad (6.1)$$

where $\mathbf{r} = (x_1, x_2, x_3)$ and the charge experiences an electric field

$$\mathbf{E} = (E_1, E_2, E_3) \exp(-i\omega t).† \qquad (6.2)$$

† Of course, an electric field must be real, and so of the form

$$\mathbf{E} = (E_1, E_2, E_3) \exp(-i\omega t) + (E_1^*, E_2^*, E_3^*) \exp(i\omega t). \qquad (6.2a)$$

However, it is simpler to analyze the problem initially using Eq. (2), and subsequently to introduce the correct form, Eq. (2a), when we take a sum over frequencies. This sum will then include negative as well as positive frequencies.

Equation (1) is, of course, just a generalization of Eq. (5.1) to a three-dimensional problem. For convenience, we assume that the damping coefficient a is a scalar, although this is not essential. We now choose a system of axes in which K_{jl} is diagonal, and denote its diagonal elements by ω_j^2, so that Eq. (1) becomes

$$\ddot{x}_j + a\dot{x}_j + \omega_j^2 x_j + \sum_j \sum_n c_{jln} x_l x_n = \frac{q}{m} E_j \exp(-i\omega t). \qquad (6.3)$$

In order to solve Eq. (3), we should like to treat the nonlinear terms $c_{jln} x_l x_n$ as a perturbation. Such a procedure will usually be justified if these terms are much smaller in magnitude than the linear terms $\omega_j^2 x_j$. Although Eq. (3) for the case when all the nonlinear terms are zero can readily be solved exactly, it is instructive to study it by dimensional analysis. We assume that the damping and nonlinear terms are small and apply dimensional analysis to Eq. (3) neglecting these terms. As we see below, the damping can only be ignored if ω is not too close to a resonant frequency ω_j. Whatever system of electrical units one chooses, $q\mathbf{E}$ has the dimensions of force, so that $qE_j/mx\omega_j^2$ is dimensionless, while ω/ω_j is obviously dimensionless. Thus

$$x_j \exp(i\omega t) = [qE_j/m\omega_j^2] f(\omega/\omega_j). \qquad (6.4)$$

We now use this result to examine the conditions under which the nonlinear terms will in fact be small. If $f(\omega/\omega_j)$ is of order unity or less, the nonlinear terms can be treated as a perturbation provided that

$$cx_j^2/\omega_j^2 x_j \approx cqE_j/m\omega_j^4 \ll 1,$$

where c is a typical nonlinear coefficient. This condition is usually found to be satisfied. If we treat the nonlinear terms as a perturbation, the zeroth-order approximation to the solution of Eq. (3) is

$$x_j^{(0)} = (q/m)E_j \exp(-i\omega t) D_j^{-1}(\omega), \qquad (6.5)$$

where

$$D_j(\omega) = \omega_j^2 - \omega^2 - ia\omega. \qquad (6.6)$$

In writing the solution in this form, we have ignored all transient effects, since we are interested only in the forced oscillations; thus, in contrast to the problems considered in Section 5.1, initial (or boundary) conditions are irrelevant. On comparing these last two equations, for $a = 0$, with Eq. (4), we find that there $f(\omega/\omega_j) = 1/(1 - \omega^2/\omega_j^2)$, and so can only be of order of magnitude greater than unity if ω is sufficiently close to ω_j, i.e., if the forcing frequency is close to one of the system's resonant frequencies.

According to the standard methods of perturbation theory, the first-order correction to this solution, which we denote by $x_j^{(1)}$, satisfies the equation

$$\ddot{x}_j^{(1)} + a\dot{x}_j^{(1)} + \omega_j^2 x_j^{(1)} = -\sum_l \sum_n c_{jln} x_l^{(0)} x_n^{(0)}. \qquad (6.7)$$

Thus,

$$x_j^{(1)} = -\frac{q^2}{m^2} \sum_l \sum_n \frac{c_{jln}}{D_j(2\omega)D_l(\omega)D_n(\omega)} E_l E_n \exp(-2i\omega t). \qquad (6.8)$$

In principle, it is possible to continue our iteration procedure, and calculate contributions to x_j to higher order in the perturbation. However, our Eq. (1) for the anharmonic oscillator arises from the neglect of all but the first three terms in the expansion of the particle's potential energy

$$V(\mathbf{r}) = V(0) + \frac{1}{2} \sum_j \sum_l K_{jl} x_j x_l + \frac{1}{3} \sum_j \sum_l \sum_n c_{jln} x_j x_l x_n$$

$$+ \frac{1}{4} \sum_j \sum_l \sum_n \sum_p e_{jlnp} x_j x_l x_n x_p + \cdots. \qquad (6.9)$$

Thus, it is inconsistent to calculate the second-order correction $x_j^{(2)}$ by using terms such as $c_{jln} x_l^{(1)} x_n^{(0)}$ while ignoring the terms $e_{jlnp} x_l^{(0)} x_n^{(0)} x_p^{(0)}$ which would appear at this stage if the effect of the fourth term in the expansion of $V(\mathbf{r})$ had been included in the oscillator's equation of motion, Eq. (1). This is a very general sort of consideration. An approximate equation should never be solved to a higher order in the small parameter than that for which the equation was derived.

The polarization induced in the medium by the field $\mathbf{E}$ is just $\mathbf{P} = Nq\mathbf{r}$, where N is the density of particles per unit volume, so that

$$P_j = Nq x_j^{(0)} + Nq x_j^{(1)} + \cdots. \qquad (6.10)$$

The first term on the right-hand side of Eq. (10) is linear in the field, and so can be incorporated in the dielectric constant of the medium. The second term gives rise to a nonlinear polarization, whose components we denote by P_j^{NL}. If the field $\mathbf{E}$ consists of several frequencies

$$\mathbf{E} = \sum_\alpha (E_1^\alpha, E_2^\alpha, E_3^\alpha) \exp(-i\omega_\alpha t), \qquad (6.11)$$

the generalization of Eq. (8) leads, in conjunction with Eq. (10), to the formula

$$P_j^{\text{NL}} = -\frac{Nq^3}{m^2} \sum_\alpha \sum_\beta \sum_l \sum_n \frac{c_{jln} E_l^\alpha E_n^\beta}{D_j(\omega_\alpha + \omega_\beta)D_l(\omega_\alpha)D_n(\omega_\beta)} \exp[-i(\omega_\alpha + \omega_\beta)t].$$

$$(6.12)$$

For the rest of this chapter, we restrict our attention to nonlinear terms in the polarization that are of second order in the field, such as those appearing in Eq. (12). These will usually be the dominant terms in $\mathbf{P}^{NL}$ provided that they do not all vanish.

The question of whether or not these second-order terms vanish depends on the symmetry of the system. For our model, according to Eq. (12), these terms will all vanish if and only if all the coefficients c_{jln} do so. From Eq. (9), we see that these coefficients have the intrinsic symmetry of being unaffected by any permutation of the indices. In addition, since $V(\mathbf{r})$ must have the full symmetry of the system, the coefficients c_{jln} must be invariant under all the system's symmetry operations.

The Model's Limitations

One serious limitation of the model that we have described is that it expresses the polarization $\mathbf{P}$ in terms of the local field $\mathbf{E}_{loc}$ experienced by a typical particle, rather than in terms of the external field $\mathbf{E}_{ext}$ which is involved in Maxwell's equations. These two fields differ because each particle is subject not only to the external field but also to the fields produced by all the other induced dipoles. For static fields inducing a linear polarization, it is usually found that

$$\mathbf{E}_{loc} = \mathbf{E}_{ext} + b\mathbf{P}, \qquad (6.13)$$

where b is the Lorentz local field factor, and equal $4\pi/3$ for systems with cubic symmetry, for instance. The generalization of this formula to nonzero frequencies and nonlinear susceptibilities is possible, but leads to much more complicated formulas than Eq. (12) for the relationship between $\mathbf{P}^{NL}$ and the external field.

Another important limitation of the model, and one of a more fundamental nature, is that it assumes that the polarization depends only on the instantaneous value of the electric field. In practice, as we discussed in Sections 4.1 and 4.3, most systems have a finite response time, in which case the polarization depends also on the value of the field at earlier times.

Thus, while the model is useful for giving a general idea of the nonlinear optical properties of systems, it cannot be used to derive general, rigorous results. This is, of course, the fate of all simple models. It does not detract from their usefulness provided that their inherent limitations are kept in mind.

6.3 Nonlinear Susceptibilities

Nonlinear Response Functions

We now wish to obtain a general formula for the nonlinear susceptibility, i.e., the relationship between $\mathbf{P}^{\mathrm{NL}}$ and the external field $\mathbf{E}_{\mathrm{ext}}$, which is independent of any model for the system, and so is not subject to the limitations of our previous model; henceforth, we will denote this external field by $\mathbf{E}$ rather than by $\mathbf{E}_{\mathrm{ext}}$. In order to derive such a formula, we define a nonlinear response function $y_{jln}^{\mathrm{NL}}(\tau_1, \tau_2)$ according to the formula†

$$P_j^{\mathrm{NL}}(t) = \int_0^\infty d\tau_1 \int_0^\infty d\tau_2\, y_{jln}^{\mathrm{NL}}(\tau_1, \tau_2) E_l(t - \tau_1) E_n(t - \tau_2). \qquad (6.14)$$

This response function is just a generalization of the linear response function $y^{\mathrm{L}}(\tau)$, defined in Eq. (4.24), which gives rise to the linear term $\mathbf{P}^{\mathrm{L}}$ in the polarization according to the formula

$$P_j^{\mathrm{L}}(t) = \int_0^\infty d\tau\, y_{jl}^{\mathrm{L}}(\tau) E_l(t - \tau). \qquad (6.15)$$

Equations (14) and (15) are just the first two terms of an expansion of the polarization $\mathbf{P}$ in terms of the electric field $\mathbf{E}$, allowing for the possibility of delayed responses. The justification for such an analytical expansion is that there is no physical reason to expect singular behavior, while as we saw in Section 4.1 all mathematical singularities must have a physical origin. The integrals in both these equations extend from 0 to ∞, and not from $-\infty$ to ∞, because of causality, as discussed in Sections 4.1 and 4.3.

Any real electric field $\mathbf{E}(t)$ can be expressed as a Fourier transform,

$$\mathbf{E}(t) = \int_{-\infty}^\infty d\omega\, \mathbf{E}(\omega) \exp(-i\omega t), \qquad \mathbf{E}(-\omega) = \mathbf{E}^*(\omega). \qquad (6.16)$$

On substituting this form into Eq. (14), we find that

$$P_j^{\mathrm{NL}}(t) = \int_{-\infty}^\infty d\omega_1 \int_{-\infty}^\infty d\omega_2\, Y_{jln}^{\mathrm{NL}}(\omega_1, \omega_2) E_l(\omega_1) E_n(\omega_2) \exp(-i(\omega_1 + \omega_2)t), \qquad (6.17)$$

where

$$Y_{jln}^{\mathrm{NL}}(\omega_1, \omega_2) = \int_0^\infty d\tau_1 \int_0^\infty d\tau_2\, y_{jln}^{\mathrm{NL}}(\tau_1, \tau_2) \exp[i(\omega_1\tau_1 + \omega_2\tau_2)]. \qquad (6.18)$$

---

† In this section, summation is implied over the letters j, l, and n whenever they appear twice in a term, in accordance with the usual summation convention.

If only fields of discrete frequencies ω_α are involved in the system, so that the total electric field is given by Eq. (11), the integrals over frequencies in Eq. (17) are replaced by sums, and we can write

$$P_j^{NL}(t) = \sum_\alpha \sum_\beta Y_{jln}^{NL}(\omega_\alpha, \omega_\beta)E_l^\alpha E_n^\beta \exp(-i(\omega_\alpha + \omega_\beta)t). \qquad (6.19)$$

Equation (12) is just a special case of Eq. (19), with a specific form for the coefficient Y_{jln}^{NL}. Since y_{jln}^{NL} is real,

$$Y_{jln}^{NL}(-\omega_1 - \omega_2) = Y_{jln}^{NL}(\omega_1, \omega_2)^*. \qquad (6.20)$$

The results contained in Eqs. (17)–(20) all follow from the general principles of analyticity and causality, and so are not subject to the limitations of our previous model. We will now consider the application of another general concept, namely, symmetry.

Free Energy and Intrinsic Symmetries

In order to derive the intrinsic symmetry of the nonlinear susceptibilities, it is most convenient to consider that part of the free energy density that is associated with the electric field. In general, the free energy density contains a term

$$\Phi_E = -\mathbf{E} \cdot \mathbf{P}, \qquad (6.21)$$

so that we can associate with the nonlinear part of the polarizability a term

$$\Phi_E^{NL} = -\mathbf{E} \cdot \mathbf{P}^{NL} = -\sum_\alpha \sum_\beta \sum_\gamma E_j^\gamma Y_{jln}(\omega_\alpha, \omega_\beta)E_l^\alpha E_n^\beta \exp(-i(\omega_\alpha + \omega_\beta + \omega_\gamma)t),$$
$$(6.22)$$

where we use Eq. (11) for $\mathbf{E}$ and Eq. (19) for $\mathbf{P}^{NL}$.

For our purposes, it is not the instantaneous value of the free energy density that interests us but rather its time average, since frequency doubling requires the concept of a finite time interval in which frequency can be defined. The only terms that contribute to the time average of Φ_E^{NL} are those with $\omega_\gamma = -(\omega_\alpha + \omega_\beta)$, so that this time average (denoted by $\bar{\Phi}$) is of the form

$$\bar{\Phi}_E^{NL} = -\sum_\alpha \sum_\beta \chi_{jln}(-(\omega_\alpha + \omega_\beta), \omega_\alpha, \omega_\beta)E_j^{-(\alpha+\beta)}E_l^\alpha E_n^\beta, \qquad (6.23)$$

where the indices j, l, n are associated, respectively, with the first, second, and third frequencies that appear in the argument of χ, and we have written, for convenience,

$$Y_{jln}(\omega_\alpha, \omega_\beta) \equiv \chi_{jln}(-(\omega_\alpha + \omega_\beta), \omega_\alpha, \omega_\beta). \qquad (6.24)$$

From Eq. (22), it is immediately obvious that χ is unaffected by a simultaneous interchange of indices and frequencies, so that, for instance

$$\chi_{njl}(\omega_3, \omega_1, \omega_2) = \chi_{jln}(\omega_1, \omega_2, \omega_3). \tag{6.25}$$

In addition, we find from Eqs. (18), (20), and (24) that

$$\chi_{jln}(-\omega_1, -\omega_2, -\omega_3) = \chi_{jln}^*(\omega_1, \omega_2, \omega_3). \tag{6.26}$$

Moreover, since the electric field is a polar vector, χ_{jln}^* must be a polar tensor. Hence, as we discussed in Section 3.4, it will be zero, for instance, in any system for which inversion is a symmetry operation. Thus, general symmetry considerations enable us to predict that effects of second order in the field will not occur for a large range of systems.

Second Harmonic Generation in KDP

As a specific example of a nonlinear optical effect, we now consider second harmonic generation, i.e., the formation of a wave of frequency 2ω from an incident wave of frequency ω. Thus, a typical nonlinear susceptibility that interests us is $\chi_{jln}(-2\omega, \omega, \omega)$. In view of the intrinsic symmetry, we find that in this case, from Eq. (25)

$$\chi_{jln}(-2\omega, \omega, \omega) = \chi_{jnl}(-2\omega, \omega, \omega), \tag{6.27}$$

i.e., this tensor is symmetric with respect to the interchange of its last two indices, and so contains a maximum of 18 distinct elements. One of the crystals often used for second harmonic generation is potassium dihydrogen phosphate, commonly known as KDP. The point group of this crystal is denoted by $\bar{4}\,2m$ or D_{2d}, and consists of the eight operations listed in Table 1, where the axis of maximum symmetry is chosen as the z axis.

We now apply the considerations of Sections 3.1 and 3.4, namely, that χ_{jln} must be invariant under any change of coordinates that results from a symmetry operation. The application of A_2 requires that the components of χ with (jln) = (111), (112), (122), (133), (211), (212), (222), (233), (313), and

Table 1

Operations of D_{2d} and Their Effects on the Coordinates

| Operation label | A_1 | A_2 | A_3 | A_4 | A_5 | A_6 | A_7 | A_8 |
Type	E	C_2	S_4	S_4	σ_d	σ_d	C_2'	C_2'
Effect on x	x	$-x$	y	$-y$	y	$-y$	$-x$	x
Effect on y	y	$-y$	$-x$	x	x	$-x$	y	$-y$
Effect on z	z	z	$-z$	$-z$	z	z	$-z$	$-z$

(323) all transform into their negatives and so vanish. Similarly, A_8 leads to the vanishing of the components with $(jln) = (113)$, (223), (311), (322), and (333). Thus the only nonvanishing components of χ in KDP are χ_{123} and χ_{213}, which are equal since A_3 converts one into the other, and χ_{312}, plus those obtained from them by interchanging the last two indices, which are equal to the original ones.

In our subsequent analysis, we also require the form of the dielectric tensor ε_{jl}, which incorporates the effect of that part of **P** that is linear in the field. The operations A_7 and A_8 lead to the vanishing of its off-diagonal elements, while A_3, for instance, leads to the requirement that $\varepsilon_{11} = \varepsilon_{22}$.

6.4   Use of Perturbation Theory

Preparation of the Problem for Perturbation Theory

We have now derived as much information as we can simply obtain from the model and from general considerations such as analyticity and symmetry. In order to proceed further and find the fields of the waves of different frequencies, we must write down and attempt to solve the equations for these fields and use perturbation theory. As we saw in Chapter 5, it is often necessary to prepare the problem in a suitable form before perturbation theory can be applied. All these preliminary steps, such as identification of the most important part of the perturbation, choice of the small parameter, and, if necessary, adjustment of the ground state will now be applied to our problem.

With our system of coordinates, in which the dielectric tensor ε_{jk} is diagonal, we readily find from Maxwell's equations that each component of an electric field of frequency ω satisfies the equation

$$\nabla^2 E_j - (\partial/\partial x_j)(\nabla \cdot \mathbf{E}) + (\varepsilon_{jj}\omega^2/c^2)E_j = -(4\pi\omega^2/c^2)P_j^{\mathrm{NL}}. \qquad (6.28)$$

We look for solutions of this equation that are of the form of plane waves advancing in the direction $\hat{\mathbf{u}}$, and denote the component of **r** in this direction by u, so that

$$u = \hat{\mathbf{u}} \cdot \mathbf{r}, \qquad (6.29)$$

and we can write

$$\mathbf{E}(\mathbf{r}, t) = \mathbf{E}^\omega(u) \exp(-i\omega t). \qquad (6.30)$$

In view of the form of P_j^{NL}, which in our approximation [Eq. (19)] involves second powers of the field amplitudes, we cannot hope to solve Eq. (28) exactly. Our first step is to find the leading terms in P_j^{NL} by considering the zeroth-order approximation to the solution of Eq. (28). In this approximation, we ignore the term P_j^{NL} on the right-hand side, and so have a homogene-ous linear equation. For a medium in which $\varepsilon_{11} = \varepsilon_{22}$, such as KDP, one solution of this equation is always a transverse wave with its electric field in the xy plane and a refractive index $\sqrt{\varepsilon_{11}}$ that is independent of the direction of $\mathbf{u}$; this solution is called the ordinary ray, and will be denoted by the subscript o. The other solution, known as the extraordinary ray and de-noted by the subscript e, is in general not transverse and has a phase velocity that depends on $\hat{\mathbf{u}}$. An exception arises if $\hat{\mathbf{u}}$ lies in the xy plane, in which case the extraordinary ray has an electric field in the z-direction and a fixed phase velocity, and for the sake of simplicity we restrict our attention to this case. If we write

$$\hat{\mathbf{u}} = (-\cos\theta, \sin\theta, 0), \tag{6.31}$$

we then readily find that the field amplitudes for the ordinary and ex-traordinary wave are, respectively, of the form

$$\mathbf{E}_o^\omega(u) = A_o \exp(ik_o u)(\sin\theta, \cos\theta, 0), \qquad k_o = (\omega/c)\sqrt{\varepsilon_{11}(\omega)} \tag{6.32a}$$

and

$$\mathbf{E}_e^\omega(u) = A_e \exp(ik_e u)(0, 0, 1), \qquad k_e = (\omega/c)\sqrt{\varepsilon_{33}(\omega)}. \tag{6.32b}$$

As we can see from Eq. (19), waves of the above form will give rise in $\mathbf{P}^{\mathrm{NL}}$ to terms of frequency 2ω that depend on u as $\exp(2iku)$, where k denotes k_o or k_e. These terms will act in Eq. (28) as a driving force that produces waves of frequency 2ω, which according to Eqs. (32) depend on u as

$$\exp[(2i\omega/c)\sqrt{\varepsilon_{jj}(2\omega)} \cdot u],$$

with $j = 1$ or 3. An appreciable amplitude at frequency 2ω will only be produced if this wave and $\mathbf{P}^{\mathrm{NL}}$ advance in the direction $\hat{\mathbf{u}}$ with nearly the same phase, and the more nearly equal are their phases, the greater will their amplitudes be. For simplicity, we consider here only the case of exactly equal phases, which is known as perfect phase matching. If there were no disper-sion whatsoever, so that ε was independent of frequency, such matching would occur between ordinary rays at frequencies ω and 2ω, and also be-tween extraordinary rays at these frequencies. In practice, however, because of dispersion, perfect phase matching for a wave advancing in the xy plane is only possible at certain frequencies, and between an ordinary wave at one frequency and an extraordinary one at the other. In order to treat a definite problem, we consider a situation in which perfect phase matching occurs only between an ordinary ray at frequency ω and an extraordinary ray at frequency 2ω, so that

$$\varepsilon_{33}(2\omega) = \varepsilon_{11}(\omega). \tag{6.33}$$

Since these phase-matched rays give rise to the dominant terms in $\mathbf{P}^{NL}$, we will ignore (even in approximations of the first and higher orders) the components of fields of frequency ω in the direction z and of the field of frequency 2ω in the xy plane. We henceforth use the subscripts o and e to refer, respectively, to the ordinary ray of frequency ω and the extraordinary ray of frequency 2ω, and so omit these frequencies as superscripts. Thus, we write

$$\mathbf{E}_o(u) = E_o(u)(\sin\theta, \cos\theta, 0), \tag{6.34a}$$

$$\mathbf{E}_e(u) = E_e(u)(0, 0, 1). \tag{6.34b}$$

In our analysis of the symmetry of $\chi_{jln}(-2\omega, \omega, \omega)$ in KDP, we found that the only nonzero elements were those with $(jln) = (123)$, (213), (312), plus those obtained from these by permuting the indices and frequencies. Since we are considering a case in which the field of frequency 2ω has components only in the z-direction, the elements of χ with $(jln) = (123)$ or (213) will not contribute to $\mathbf{P}^{NL}$. It is convenient to write

$$\chi^*_{132}(-\omega, 2\omega, -\omega) = \chi_{312}(-2\omega, \omega, \omega) = 4\mu, \tag{6.35}$$

where the first equality follows from the intrinsic symmetry, Eq. (26). On substituting in Eq. (19) for $P_j^{NL}(t)$ from Eqs. (34) and (35), and taking real parts of the field, we then readily find that

$$P_1^{NL} = \mu^* E_o^* E_e \cos\theta \cdot \exp(ik_o u - i\omega t) + \text{complex conjugate}, \tag{6.36a}$$

$$P_2^{NL} = \mu^* E_o^* E_e \sin\theta \cdot \exp(ik_o u - i\omega t) + \text{complex conjugate}, \tag{6.36b}$$

$$P_3^{NL} = \mu E_o^2 \exp(ik_e u - 2i\omega t) + \text{complex conjugate}, \tag{6.36c}$$

while for perfect phase matching

$$k_e = 2k_o. \tag{6.37}$$

Hence, on substituting in Eq. (28) and taking components in the directions of the fields of Eqs. (34), we find that, if μ is real†

$$d^2E_o/du^2 + k_o^2 E_o = -(4\pi\omega^2/c^2)(2\mu \sin 2\theta)E_e E_o^*, \tag{6.38a}$$

and

$$d^2E_e/du^2 + k_e^2 E_e = -(4\pi(2\omega)^2/c^2)(\mu \sin 2\theta)E_o^2. \tag{6.38b}$$

---

† This is a reasonable approximation provided that ω and 2ω are not too close to a resonant frequency, since then the imaginary part of μ is much smaller than its real part. For instance, for the model system described by Eqs. (3) and (12), if $a \ll \omega_j$ as we have assumed then the ratio of the imaginary part of μ to its real part is approximately $2a\omega/[(\omega_3^2 - 4\omega^2)(\omega_1^2 - \omega^2)]$.

Application of Perturbation Theory

Our problem is now almost ready for the application of perturbation theory. The last step before we do this is to introduce the dimensionless variable

$$v = k_0 u. \tag{6.39}$$

In terms of this, Eqs. (38) can be written

$$d^2 E_o/dv^2 + E_o + 2\gamma E_e E_o^* = 0, \tag{6.40a}$$

$$d^2 E_e/dv^2 + 4E_e + 4\gamma E_o^2 = 0, \tag{6.40b}$$

where, in view of the definition of k_o in Eq. (32a),

$$\gamma = 4\pi\mu \sin(2\theta)/\varepsilon_{11}(\omega). \tag{6.41}$$

In principle, Eqs. (40) can be solved by expanding E_o and E_e, or more usefully $E_o \exp(-iv)$ and $E_e \exp(-2iv)$, as power series in γ. However, just as for the anharmonic oscillator in Section 5.1, such a procedure can lead to problems of divergence. Instead, just as in that case, we assume that the solutions of our equations are a modification of those for $\gamma = 0$. In Section 5.1, we found that the first approximation to the solution contained a term of the form $\cos(\omega_o t + \gamma\omega_1 t)$, where ω_1 was a constant. In the present case, we assume a similar sort of behavior, namely, that the amplitudes and phases of the waves are functions of γv (and so are constants for $\gamma = 0$, of course). Thus we write

$$E_o(v) = A(\gamma v) \exp[iv + ia(\gamma v)], \qquad A \text{ and } a \text{ real functions,} \quad (6.42a)$$

$$E_e(v) = B(\gamma v) \exp[2iv + ib(\gamma v)], \qquad B \text{ and } b \text{ real functions.} \quad (6.42b)$$

In other words, we are using a renormalized phase, a procedure similar to the renormalized frequency used in Section 5.1.

On substituting these expressions in Eqs. (40), retaining only terms of first order in γ, and writing $w = \gamma v$, we then readily find that

$$i \, dA/dw - A \, da/dw + AB \exp(ib - 2ia) = 0, \tag{6.43a}$$

$$i \, dB/dw - B \, db/dw + A^2 \exp(2ia - ib) = 0. \tag{6.43b}$$

We now write

$$\psi = 2a - b \tag{6.44}$$

and take real and imaginary parts of Eqs. (43). From these, we find that

$$dA/dw = AB \sin \psi, \tag{6.45a}$$

$$dB/dw = -A^2 \sin \psi, \tag{6.45b}$$

and

$$d\psi/dw = (2B - A^2/B)\cos\psi. \tag{6.46}$$

Incidentally, on multiplying Eqs. (45a) and (45b) by A and B, respectively, and adding, we find that

$$(d/dw)(A^2 + B^2) = 0, \tag{6.47}$$

so that

$$A^2 + B^2 = C^2, \tag{6.48}$$

where C is some constant. This is just an expression of the conservation of energy.

In order to solve this set of equations, we note that, from Eqs. (45),

$$(d/dw)(\ln A^2 B) = (2B - A^2/B)\sin\psi, \tag{6.49}$$

so that, on substituting in Eq. (46) and integrating, we find that

$$A^2 B \cos\psi = \text{const.} \tag{6.50}$$

A particularly simple case, which we now consider, is when only a wave of frequency ω is incident on the crystal at the surface $u = 0$. In that case, $B(0) = 0$, so that in view of (50)

$$A^2 B \cos\psi = 0, \tag{6.51}$$

while, in order that a wave of frequency 2ω be generated we require that $B'(0) > 0$. These two conditions will only be satisfied if $\psi = -\pi/2 + 2r\pi$ for some integer r. In that case, on substituting this value of ψ and $A^2 = C^2 - B^2$ from Eq. (48) into Eq. (45b), we find on integrating and using our boundary condition at $w = 0$ that

$$B(w) = C\tanh(Cw), \tag{6.52}$$

and hence, from Eq. (48), that

$$A(w) = C\,\text{sech}(Cw), \tag{6.53}$$

while in view of Eqs. (43) and (51) a and b are constants. Finally we substitute these results in Eqs. (42), choose $a(0) = 0$, and combine the frequencies $\pm\omega$, $\pm2\omega$, to obtain real waves. We then find that, to this level of approximation, the fields of the wave in the xy plane and parallel to the z axis are, respectively, of magnitude

$$E_{xy}(u, t) = 2C\,\text{sech}(\gamma C k_o u)\cos(k_o u - \omega t), \tag{6.53a}$$

$$E_z(u, t) = 2C\tanh(\gamma C k_o u)\sin(2k_o u - 2\omega t). \tag{6.53b}$$

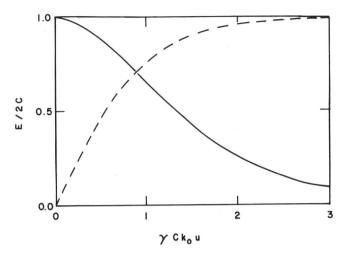

Fig. 1. The amplitude E of the electric fields of the incident wave (full line) and of the second harmonic wave that is generated (broken line), as a function of the distance from the edge of the crystal.

In Fig. 1, we show the amplitudes of these fields as a function of u. For a crystal of finite length L, in which virtually all the wave incident at the surface $u = L$ is transmitted, the ratios of the intensities of the transmitted waves of frequency 2ω and ω will be just $E_z(L, t)^2/E_{xy}(L, t)^2$, i.e., $\sinh^2(\gamma Ck_o L)$, which increases rapidly with L. Moreover, for small L such that $\gamma Ck_o L \ll 1$, the intensity at frequency 2ω is proportional to the square of the length of the crystal, and to the square of the intensity of the incident wave of frequency ω.

Conclusions

In this problem, we seem to have obtained a specific analytical form for the solution, and not just an estimate of its magnitude as was our declared aim. However, this solution is only correct to first order in the small parameter γ or μ. Moreover, in deriving it we have made a lot of simplifying assumptions. For instance, we have assumed perfect phase matching, i.e., that the waves of frequency ω and 2ω have exactly the same phase velocity. If this does not occur, our solution will not be accurate, but we expect that it should be a good approximation (at least to order of magnitude) provided that the difference in velocities produces, over the whole length of the crystal, a phase difference that is much less than π. We have also assumed that $B(0) = 0$, i.e., that there is no wave of frequency 2ω at the front surface of the crystal, and have ignored the problems of transmission and reflection of light at the crystal's other surface. Additional complications will arise if we consider waves not advancing in a symmetry plane, or crystals of lower symmetry. It is, in fact, possible to find for some of these cases solutions as accurate as our Eq. (53) was for our simple model, but these solutions are of a much more complicated form. However, it is reasonable to expect that our simple solution will in most cases give correctly the order of magnitude of the exact solution, and this is what we aim to achieve by the qualitative analysis of physical problems.

References

Chapter 1

§1.1 A good discussion of the equations of state of gases is contained in

[1] K. Huang, "Statistical Mechanics," Chapter 15. Wiley, New York, 1963.

The case that is well described by van der Waals equation is discussed in

[2] M. Kac, G. E. Uhlenbeck and P. C. Hemmer, *J. Math. Phys.* **4**, 216 (1963).

§1.2 An excellent elementary introduction to models of the atomic nucleus is contained in the review article by

[3] R. Van Wageningen, *Am. J. Phys.* **28**, 425 (1960).

A more detailed treatment of these models is contained, for instance, in the book

[4] I. Kaplan, "Nuclear Physics." Addison-Wesley, Reading, Massachusetts, 1963.

More recent developments are described fairly simply by

[5] G. E. Brown, "Unified Theory of Nuclear Models and Forces." North-Holland Publ., Amsterdam, 1971.

§1.3 Simple reviews of the quark model are presented in

[6] R. O. Young, *Am. J. Phys.* **41**, 472 (1973).

and

[7] S. L. Clashow, *Sci. Am.* **233**(4), 38 (October, 1975).

A review of more recent developments is contained in

[8] F. E. Close, *Rep. Prog. Phys.* **42**, 1285 (1979).

§1.4 A detailed description of different types of quasi-particles is contained in the book

[9] D. Pines, "Elementary Excitations in Solids." Benjamin, New York, 1964.

A history of how they were given their names is presented in

[10] C. T. Walker and G. A. Slack, *Am. J. Phys.* **38**, 1380 (1970).

§1.5 The simplified theory of space-charge-limited currents was proposed and clearly described in

[11] M. A. Lampert, *Phys. Rev.* **103**, 1648 (1956).

A good account of the theory, including subsequent developments, is presented in

[12] D. R. Lamb, "Electrical Conduction Mechanisms in Thin Insulating Films," Chapter 4. Methuen, London, 1967.

§1.6 An excellent and comprehensive textbook on hydrodynamics is

[13] L. D. Landau and E. M. Lifshitz, "Fluid Mechanics." Pergamon Oxford, 1959,

in which Chapter IV is devoted to boundary layers.

A classical text on boundary layer theory, which considers turbulent as well as laminar flow, is

[14] H. Schlichting, "Boundary-Layer Theory," 6th ed. McGraw-Hill, New York, 1968.

Chapter 2

§2.1 A good account of the selection of units and different systems of unity is presented in

[15] B. S. Massey, "Units, Dimensional Analysis and Physical Similarity." Van Nostrand-Reinhold, Princeton, New Jersey, 1971,

which also contains an account of dimensional analysis. The classical text in this field is that of

[16] P. W. Bridgman, "Dimensional Analysis." AMS Press, New York, 1978.

The scaling theory of phase transitions, and its experimental verification, is described in the book

[17]   H. Stanley, "Introduction to Phase Transitions and Critical Phenomena." Oxford Univ. Press (Clarendon), London and New York, 1971,

as well as in Ref. [23].

§2.2   The use of vector lengths is discussed in

[18]   H. E. Huntley, "Dimensional Analysis." Dover, New York, 1967.

§2.3   There are a large number of good textbooks in statistical physics, such as Ref. [1]

and

[19]   L. D. Landau and E. M. Lifshitz, "Statistical Physics," 2nd ed. Pergamon, Oxford, 1969.

§2.4   A good account of the derivation of Thomas–Fermi equation and its application to ions as well as to neutral atoms is contained in

[20]   M. Born, "Atomic Physics," 8th ed., Chapter VI §9. Blackie, Glasgow and London, 1969.

A more recent review of the theory and its possible generalisations is presented in

[21]   D. A. Kirzhnitz, Yu. E. Lozovik and G. V. Shpatankavskaya, *Sov. Phys. Usp.* **18**, 649 (1975).

A brief account of physical similarity is contained in Chapter 7 of Ref. [15]; in particular, Section 7.4 of this book contains a comprehensive list of the named dimensionless parameters and their definitions.

An interesting and systematic, though not very simple, book that discusses dimensional analysis, dimensionless equations and both exact and approximate physical similarity, and their application to the practical solution of problems, is

[22]   S. J. Kline, "Similitude and Approximation Theory." McGraw–Hill, New York, 1965.

§2.5   Some recent results of renormalisation group theory can be found in the book

[23]   S. K. Ma, "Modern Theory of Critical Phenomena." Benjamin, New York, 1976.

The results of computer calculations for the two-dimensional Ising model on a triangular lattice (which, incidentally, do not converge in the third approximation) are reported in

[24]   P. C. Hemmer and M. G. Verlarde, *J. Phys. A.* **9**, 1713 (1976).

Similar calculations for the square lattice are reported in

[25]   S. C. Hsu, Th. Niemejer, and J. D. Guorton, *Phys. Rev. B* **11**, 2699 (1975).

Chapter 3

§3.1   Of the numerous textbooks on group theory, one of the simplest to read and use is

[26]   J. W. Leech and P. J. Newman, "How to Use Groups." Methuen, London, 1969.

A standard textbook on the number of nonvanishing components in a tensor is

[27]   J. F. Nye, "Physical Properties of Crystals." Oxford Univ. Press (Clarendon), London and New York, 1957.

An excellent more recent book, which also includes a discussion of magnetic groups and of Onsager's principle, is

[28]   S. Bhagavantam, "Crystal Symmetry and Physical Properties." Academic Press, New York, 1966.

The order–disorder theory of phase transitions is discussed in Ref. [19, Chapter XIV].

§3.2   While all the necessary information about quantum mechanics is contained in the text, more details can be found in any of the numerous textbooks on quantum mechanics, such as

[29]   L. Schiff, "Quantum Mechanics," 3rd ed. McGraw–Hill, New York, 1968.

§3.3   Most of the topics treated in the first part of this section are considered in Ref. [26]. The irreducible representations of crystals, and their application to phase transitions, are considered in Ref. [19, Chapters XIII and XIV]. A modification of this theory of phase transitions is presented by

[30]   A. Michelson, *Phys. Rev. B* **18**, 459 (1978).

Special points for integration over the first Billouin zone are defined and derived by

[31]   D. J. Chadi and M. L. Cohen, *Phys. Rev. B* **8**, 5747 (1973).

   §3.4   A simple introduction to magnetic groups is presented in

[32]   A. P. Cracknell, *Contemp. Phys.* **5**, 459 (1967).

A much more detailed account is contained in the book

[33]   R. R. Birss, "Symmetry and Magnetism." North-Holland Publ., Amsterdam, 1966.

Chapter 4

   §4.1   The behavior of a three-component plasma of charged particles is investigated in

[34]   M. Gitterman and V. Steinberg, *J. Chem. Phys.* **69**, 2763 (1978), and *Phys. Rev. A* **20**, 1236 (1979).

Singularities in quantum mechanics are discussed in

[35]   L. D. Landau and E. M. Lifshitz, "Quantum Mechanics," 2nd ed., Sections 32 and 35. Pergamon, Oxford, 1965.

These authors, in Chapter XVII, as well as Ref. [29, Chapter 9], discuss scattering problems and some dispersion relations associated with them.
   A much more detailed treatment of these topics is contained in the book

[36]   H. M. Nussenzveig, "Causality and Dispersion Relations." Academic Press, New York, 1972.

A good discussion of dispersion relations and sum rules for macroscopic quantities such as the dielectric constant is contained in

[37]   F. Stern, Elementary Theory of the Optical Properties of Solids, *in* "Solid State Physics," Vol. 15, p. 300. F. Seitz and O. Turnbull (eds.) Academic Press, New York, 1963.

Fluctuations, Brownian motion, and related topics are presented fairly simply in Part 2 of the book

[38]   C. Kittel, "Elementary Statistical Physics." Wiley, New York, 1958.

An observed divergence of the diffusion coefficient as $1/t$ for two-dimensional systems has been discussed by

[39]   S. Toxvaerd, *Phys. Rev. Lett.* **43**, 529 (1979).

§4.2   See references for Section 4.1.

The application of scattering matrix theory to interactions between nucleous and elementary particle physics is considered in the following two books:

[40]   R. J. Eden, P. V. Landshoff, D. I. Olive, and J. C. Polkinghorne, "The Analytic S-matrix." Cambridge Univ. Press, London and New York, 1966.

[41]   H. Burkhardt. "Dispersion Relation Dynamics." North-Holland Publ., Amsterdam, 1969.

§4.3   A full discussion of Titchmarch's theorem and the connection between causality and dispersion relations is contained in Ref. [36, Chapter 1].

The application of dispersion relations to the reflection of light from a thin film of material is considered briefly in Ref. [37]; the approach that we describe is that of

[42]   W. M. Hansen and W. A. Abdou, *J. Opt. Soc. Amer.* **67**, 1537 (1977).

The derivation of additional sum rules, and the physical significance of some sum rules, are studied in

[43]   M. Altarelli and D. Y. Smith, *Phys. Rev. B* **9**, 1290 (1974).

§4.4   A very clear article describing the fluctuation-dissipation theorem is

[44]   W. Bernard and H. Callen, *Rev. Mod. Phys.* **31**, 1017 (1959).

An excellent review article about time-correlation functions and transport coefficients is

[45]   R. Zwanzig, *Ann. Rev. Phys. Chem.* **16**, 67 (1965).

The generalization of the Kubo formalism to thermal disturbances is studied in

[46]   R. Kubo, M. Yokata, and S. Nakayima, *J. Phys. Soc. Jpn.* **12**, 1203 (1957).

Chapter 5

§5.1   A brief discussion of perturbation series, including the renormalization of variables, is contained in Ref. [22, Section 4.9].

The relationship between Rayleigh–Schrödinger and Brillouin–Wigner perturbation theory is discussed in detail in

[47]   W. Silvert, *Am. J. Phys.* **40**, 557 (1972)

and also, in an approach similar to that used by us, in

[48]   N. Moiseyer and J. Katriel, *Phys. Lett. A* **54**, 125 (1975).

The latter authors also discuss the use of a mixture of these two techniques. The modern kinetic theory of gases is reviewed in the article

[49]   J. R. Dorfman and H. van Beijeren, Kinetic Theory of Gases, *in* "Statistical Mechanics," Part B (B. T. Berne, ed.). Plenum, New York, 1977.

§5.2   Most of the topics in this section are treated in the standard textbooks on advanced quantum mechanics and/or applied mathematics.

A fairly simple example of the application of Dyson's equation to random systems is presented in

[50]   S. F. Edwards, *Philos. Mag.* **6**, 617 (1961).

§5.3   A full account of the Born–Oppenheimer approximation is contained in

[51]   M. Born and K. Huang, "Dynamical Theory of Crystal Lattices," Chapter IV. Oxford Univ. Press (Clarendon), London and New York, 1966.

The approach to electronic states in amorphous systems that we describe is that proposed in the last section of

[52]   D. J. Thouless, *Phys. Rep.* **13**, 93 (1974).

§5.4   The effective mass approximation is fully described in

[53]   J. M. Ziman, "Principles of the Theory of Solids," Chapter 6. Cambridge Univ. Press, London and New York, 1965.

A clear treatment of the magnetic interactions of nuclei through conduction electrons is contained in the original paper on this topic,

[54]   M. A. Ruderman and C. Kittel, *Phys. Rev.* **96**, 99 (1954).

Chapter 6

The material in this chapter is described in more detail in

[55]   P. G. Harper and B. S. Wherret (eds.), "Nonlinear Optics" (Proceedings of the Sixteenth Scottish Universities Summer School in Physics, 1975), Chapters 1 and 2. Academic Press, New York, 1977.

Index